UMAP

Models

Tools for Teaching 1994

published by

The Consortium for Mathematics
and Its Applications, Inc.
Suite 210
57 Bedford St.
Lexington, MA 02173

edited by

Paul J. Campbell
Campus Box 194
Beloit College
700 College St.
Beloit, WI 53511–5595
campbell@beloit.edu

Table of Contents

Invitation to the Celebration

Paul J. Campbell
Dept. of Mathematics and Computer Science
Beloit College
Beloit, WI 53511
campbell@beloit.edu

This special volume of *Tools for Teaching* commemorates the first ten years of the Mathematical Contest in Modeling (MCM).

The MCM came into existence amazingly quickly after Ben Fusaro had his first ideas about it in late 1983. Vague concepts became concretized in a grant proposal, the proposal was submitted and funded, a supervising committee was put in place, the contest was designed and publicized—and the first MCM took place only 16 months after Ben's first idea. In his contribution to this volume, he recounts the history, background, and emergence of the MCM.

Since that first contest, participation has grown every year, with the 1994 MCM featuring 315 teams from 192 institutions in 42 states, the District of Columbia, and 9 foreign countries.

Short articles in this volume give the current rules of the MCM, describe the process of judging the solution papers, and offer assorted short tips from advisors of teams.

This volume also contains all of the 20 problems set in the first ten years of the contest. For each year, one Outstanding paper is included, together with accompanying commentaries on all of the Outstanding papers for that problem and a brief comment from the Contest Director.

A few Outstanding papers have used highly sophisticated mathematics, such as neural nets or genetic algorithms. Often, however, simple mathematics may suffice to realize great insight. Above all, as Prof. Fusaro remarks in his account of background and history, the value of the MCM lies not in the competition but in the educational experience of students. To illustrate, we offer two case studies of institutions that do not offer advanced mathematics but have contributed teams to MCM. The first, by Alice Williams, features Southern Seminary Junior College in Virginia, which participated in the first MCM. The second case study, by Robert J. Henning, describes how he built sufficient interest in the MCM at Northcentral Technical College in Wisconsin that departments such as Radiation Technology and Auto Mechanics regularly enter teams and local industrial firms offer financial support.

Henry Ricardo describes parallels between the reform movement in college mathematics education and the MCM; modeling is central to both. Thomas O'Neil, who has coached teams in each of the contests since the first one, offers his reflections on the experience. Pat Lambert describes how the excitement builds at the University of Alaska Fairbanks—which has had several Outstanding teams—as the date of the contest approaches.

All of the participants in the MCM—and you, its spectator—are greately indebted to Ben Fusaro, whose idea it was and who served as Contest Director for the first seven years, and to Frank Giordano (U.S. Military Academy, West Point, NY), who has been Contest Director since then.

We hope that you enjoy this volume. If your institution has not yet participated in the MCM, you may obtain information and arrange to register for the next contest by contacting:

COMAP
57 Bedford St., Suite 210
Lexington, MA 02173
(800) 77–COMAP

Postscript

As this volume goes to press, the Outstanding teams from the 1995 MCM have been announced. They are:

Helix Intersection Problem:
    Iowa State University, IA
    Macalester College, MN
    North Carolina School of Science and Mathematics, NC

Salary Schedule Problem:
    Harvey Mudd College, CA (two teams)
    Southeast Missouri State University, MO
    University of Alaska Fairbanks, AK

The problem statements and the Outstanding papers will appear in *The UMAP Journal* 16 (3) late in 1995.

About the Author

Paul Campbell received a B.S. in mathematics from the University of Dayton and M.S. (algebra) and Ph.D. (mathematical logic) degrees from Cornell University. He is a professor of mathematics and computer science at Beloit College, where he served as Director of Academic Computing from 1987 to 1990. He has been the editor of *The UMAP Journal* since 1984, a responsibility that he enjoys immensely and has led him to become even more of a "generalist."

Outstanding MCM Teams 1985–1994

Beloit College, WI	1991, 1994
California Institute of Technology, CA	1989
California Polytechnic State University, CA	1989, 1990
Calvin College, MI	1985, 1987
Colorado School of Mines, CO	1986
Cornell University, NY	1993
Drake University, IA	1988, 1989
Georgetown University, DC	1986
Grinnell College, IA	1986, 1994
Harvard University, MA	1988
Harvey Mudd College, CA	1985, 1986, 1986, 1989, 1991
Hiram College, OH	1990
Humboldt State University, CA	1990
Moorhead State University, MN	1987
Mount St. Mary's College, MD	1985, 1991
Nazareth College, NY	1993
New Mexico State University, NM	1985
North Carolina School of Science and Mathematics, NC	1988, 1989, 1992, 1994
Ohio State University, OH	1989
Oklahoma State University, OK	1992
Pomona College, CA	1986, 1992
Rensselaer Polytechnic Institute, NY	1987
Ripon College, WI	1991
Rose-Hulman Institute of Technology, IN	1990
Southern Methodist University, TX	1985
Southern Oregon State University, OR	1990
U.S. Air Force Academy, CO	1990
U.S. Military Academy, NY	1988, 1993
University of Alaska Fairbanks, AK	1990, 1991, 1993
University of Calgary, Alberta, Canada	1994
University of California—Berkeley, CA	1986, 1993
University of Colorado—Boulder, CO	1992
University of Colorado—Denver, CO	1987
University of Dayton, OH	1989
University of North Carolina, NC	1994
University of Toronto, Ontario, Canada	1986, 1988, 1994
University of Western Ontario, Ontario, Canada	1991
Washington University, MO	1985, 1986, 1989, 1992

Background and History of the MCM

Bernard (Ben) A. Fusaro
Dept. of Mathematics and Computer Science
Salisbury State University
Salisbury, MD 21801
bafusaro@sae.towson.edu

Genesis

The concept of a national applied mathematics contest for undergraduates occurred to me in October 1983. The idea surfaced because of difficulties that we were having at Salisbury State University with getting our students to prepare for the Putnam Mathematical Competition.

Salisbury State has a high percentage of first-generation college students, and they tend to view facing such a formidable exam as an ordeal. The practice of the Putnam of reporting a large proportion of low numerical scores adds to the chilling effect. Finally, the small amount of applied content of Putnam problems did not generate much enthusiasm in practical-minded students.

There was much more to my notion of an applied mathematics contest than just offering different questions that would merit higher scores. For a dozen years, I had chafed at the overemphasis in established mathematics of the pure, formalistic approach, almost devoid of content. On many campuses, there was scarcely any appreciable applied or constructive mathematics presence.

In my mind, (classical) applied mathematics, computational mathematics, and statistics are as important a part of contemporary mathematical activities and curricula as pure mathematics. The model that I had in mind represents each of these four as a vertex of a tetrahedron. The edges, faces, and interior represent activities such as applied linear algebra, numerical analysis, or operations research. The Putnam deals with a small neighborhood of the pure mathematics vertex at the lofty apex of the tetrahedron. It would be difficult to tell from Putnam questions that the computer even existed.

These thoughts merged and then popped up in verbal form as "Applied Putnam." My on-campus colleagues liked the idea, but it seemed prudent to check with some off-campus mathematicians who had long involvements in applied mathematics. Calls to M.S. Klamkin (University of Alberta), H.O. Pollak (Bell Labs), and E.Y. Rodin (Washington University) elicited favorable responses and encouragement to proceed. I then called A.P. Hillman, who has had many years of experience with the Putnam. He urged me to start with a small pilot project and warned that I might be underestimating the difficulty of starting a national contest. (He was right.)

Being Chair of the Education Committee of the Society for Industrial and Applied Mathematics (SIAM) gave me a natural forum for this project. I sent an outline of a proposal for a pilot contest to the committee in November 1983. The gist of the proposal is illustrated in **Table 1**.

Table 1.
The original proposal for an "Applied Putnam."

	Pure Putnam	"Applied Putnam"
Time of contest	December	March
Sessions	Two (three hours each)	Two (three hours each)
Number of problems	12	2
Type of problems	Structural, pure	Contextual, applied
Format	Individual students No calculator or computer aids	Teams of three students Microcomputers allowed

The committee liked the proposal but had strong reservations about the time allotted per problem. The feeling was that an applied mathematics problem could not be done in half a day; estimates ran from a day to a week. One experienced SIAM officer said that a realistic problem would need a whole semester! These observations, coupled with my own fairly unshakable view that a contest for undergraduates should not occupy more than a weekend, doomed one of my favorite schemes, that teams should be required to do one continuous problem and one discrete problem.

Although the committee looked on the idea with favor, SIAM's leadership felt that the committee already had enough projects and that it should continue to concentrate on the K–12 level. However, so many people had judged the idea to be good and workable that I decided to seek another forum.

Funding

Warren Page, then editor of the *College Mathematics Journal*, gave an invited lecture to the Maryland–DC–Virginia Section of the Mathematical Association of America in November of 1983. His lively presentation included many applied examples, so I approached him after his talk. Page listened as attentively as he could, while being badgered by another mathematician who kept trying to tie Page's talk to the Vietnam War. Page's initial reaction was that the concept was interesting but not feasible.

About three weeks later, Page called me at home to say that he had given this concept of an applied contest quite a bit of thought. He had come around to the position that it was a valuable idea and that it ought to be done. Moreover, he had broached the subject to Sol Garfunkel, Executive Director of COMAP, which had been supporting applicable mathematics in a variety of ways since 1972. Garfunkel was very enthusiastic, and Page urged me to get in touch with him.

Although I was a member of COMAP and had used its materials, I had never had any interaction with Garfunkel. After one phone conversation, it was clear that we had similar goals. In fact, my personal campaign to "increase the applied mathematics presence on campus" might be one way to describe what COMAP had been doing over the years. He suggested that a proposal for a three-year grant be sent to the Fund for the Improvement of Post-Secondary Education (FIPSE) of the U.S. Dept. of Education, with COMAP the administering body and me as the Project Director. FIPSE had a reputation for backing novel ideas that might have far-reaching effects. The derivative term "Applied Putnam" was transformed into "Mathematical Competition [now Contest] in Modeling." A preliminary proposal made FIPSE's January 1984 deadline, and the three-year proposal was approved in June 1984.

Goals

The goals and purposes of the MCM are best described by two paragraphs from the abstract of the proposal to FIPSE:

> The purpose of this competition is to involve students and faculty in clarifying, analyzing, and proposing solutions to open-ended problems. We propose a structure which will encourage widespread participation and emphasize the entire modeling process. Major features include:
>
> - The selection of realistic open-ended problems chosen with the advice of working mathematicians in industry and government.
> - An extended period of time for teams to prepare solution papers within a clearly defined format.
> - The ability of participants to draw on outside resources including computers and texts.
> - An emphasis on clarity of exposition in determining final awards with the best papers published in professional mathematics journals.

As the contest becomes established in the mathematics community, new courses, workshops, and seminars will be developed to help students and faculty gain increased experience with mathematical modeling.

Organizing

Garfunkel and I were in firm agreement that the contest must be primarily an educational experience, not a competitive one. In a sense, we wanted it to be closer to the spirit of traditional English sport than to modern American sports.

I formed an advisory board of mathematical scientists who had been early backers of an applied mathematics contest:

- A.P. Hillman, University of New Mexico

- M.S. Keener, Oklahoma State University

- H.O. Pollak, Bellcore

- F.J. Roberts, Rutgers University

- E.Y. Rodin, Washington University

- L.H. Seitelman, Pratt & Whitney

- Maynard Thompson, Indiana University

- myself as chair.

Hillman, who for many years directed grading for the Putnam, agreed to be chief grader. This coup eliminated one of the two swords that hung over our heads: finding suitable problems, and judging.

The advisory board first met in August 1984. We selected two types of problems, approved ground rules, set up a Putnam-like system of faculty advisors, and established a classification for solution papers. We set the inaugural contest for the weekend of 15 February 1985. The meeting was very productive; but we departed with a note of concern over the short amount of time for publicity, registration, and final write-up of contest materials. We wondered whether we could get our predicted 55 colleges to enter the first contest.

It turned out that 158 teams, representing 104 colleges, registered for the first contest, a response that overwhelmed us. Any more than 100 solution papers would be unmanageable; there wouldn't be enough judges to allow for multiple readers for each paper. It turned out that 90 papers, representing 70 colleges, were submitted—a large but tractable number. The MCM was a success!

Acknowledgment

This history of the foundation of the MCM is adapted from the author's "Mathematical Competition in Modeling" in *Mathematical Modelling* [continued as *Mathematical and Computer Modelling*] 6 (6) (1985): 473–484. Reproduced by permission.

About the Author

Ben Fusaro has a B.A. from Swarthmore College, an M.A. in analysis from Columbia University, a Ph.D. in partial differential equations from the University of Maryland, and most recently (1990) an M.A. in computer science from the University of Maryland.

He taught at several other colleges and universities before coming to Salisbury State in 1974, where he served as chair of the Mathematics and Computer Science Dept. 1974–82 and received the Distinguished Faculty Award in 1992. Among his visiting positions, Ben was NSF Lecturer at New Mexico Highlands University and the University of Oklahoma, Fulbright Professor at National Taiwan Normal University, and visiting professor at the U.S. Military Academy at West Point. He has taught most undergraduate mathematics courses, plus graduate courses in integral equations, partial differential equations, and mathematical modeling.

In recent years, Ben has been a major exponent of environmental mathematics, a topic on which he has presented several minicourses. He is currently Project Director for an NSF grant to develop classroom materials for a modeling-based precalculus experience. For 1995–96, he will be on sabbatical at Florida State University.

Ben was the founder and the MCM and its director for the first seven years.

Ground Rules for the MCM

Registration of Teams

Teams must be registered in advance. Each department may sign up either one or two teams of three undergraduates each, but no more than four teams may be registered from any one institution. Team members may be changed at the last minute without notifying COMAP. If a department preregisters only one team but later wishes to add a second team, the department must obtain a new control number for the second team.

Departments registering for the contest will receive a contest packet that provides the current rules, requirements, and suggested procedures.

Contest Date and Time

The contest is held on a weekend in February. It begins at 12:01 A.M. local time on a Friday and ends the following Monday at 5:00 P.M. local time.

The team advisor must send to COMAP via express service the original solution paper and a copy, by a particular deadline specified in the contest materials. The paper must be typed, except for diagrams, graphs, computer programs, etc. A solution paper in another language must be followed by an English translation to arrive by a specified date.

The Advisor

The advisor is the key to the success of MCM. The advisor alerts students to the competition and its benefits and encourages the organization of teams. It is both legitimate and desirable to coach and prepare teams.

Rules

Each team receives the same two problems to consider but may submit a solution paper for only one.

A team may use any *inanimate* objects, materials, or sources of data—computers, software packages, reference works, computer network files, etc.—and should credit all sources used.

A team may not seek help in clarifying the problem or in obtaining its answer, from the team advisor or anyone else; each team is expected to develop all of its substantive analysis without the help of others. The intent of this rule is to allow participants to use all commonly available information but to disqualify any team that discusses the problems with people in a position to supply information reflecting experience or professional expertise.

Neither the school name nor team members' names may appear anywhere in the solution paper; they may appear only on the control sheet.

Suggested Outline

- A clarification or restatement of the problem, as appropriate.

- A clear exposition of all assumptions and hypotheses.

- An analysis of the problem justifying or motivating the modeling to be used.

- The design of the model.

- A discussion of how the model can be tested, including error analysis and stability (conditioning, sensitivity, etc.).

- A discussion of the strengths and weaknesses of the model.

- A one-page summary of the results, typed on the Summary Sheet, must be attached to the front of each copy of the solution paper.

Judging

Partial solutions are acceptable. There is no passing or failing score, nor are numerical scores assigned.

The MCM judges are interested primarily in the approach and methods. Conciseness and organization are extremely important. Key statements should present major ideas and results.

Each solution paper must include a one-page typed summary on the Summary Sheet included with the contest materials. *It is unlikely that MCM judges will read beyond a poorly constructed summary.*

Results, Recognition, and Prizes

Judging takes place three weeks after the contest. Solutions are recognized as Successful Participation, Honorable Mention, Meritorious, or Outstanding.

The advisors and teams are notified of the results in April. COMAP issues news releases and announcements in college and professional publications.

All successful participants receive a certificate. Outstanding teams receive bronze plaques, and their solution papers are published in a special issue of *The UMAP Journal*.

The Institute for Operations Research and the Management Sciences (INFORMS)—which resulted from the merger of the Operations Research Society of America and The Institute of Management Sciences—designates one Outstanding team from each problem as an INFORMS Winner. Each member of these teams receives a three-year membership in INFORMS and a cash prize. Each member of the other Outstanding teams recieves a one-year membership in INFORMS.

The Society for Industrial and Applied Mathematics (SIAM) designates one Outstanding team from each problem as a SIAM Winner. Each designated team receives a certificate, a cash prize, and partial expenses for a trip to the SIAM annual meeting.

Major funding and support for the MCM is provided by the U.S. National Security Agency and SIAM.

Judging the MCM

Bernard (Ben) A. Fusaro
Dept. of Mathematics and Computer Science
Salisbury State University
Salisbury, MD 21801
bafusaro@sae.towson.edu

Introduction

The process of judging the MCM has evolved somewhat through the years, to accommodate quadrupling of the number of entries in the first contest; but the essential elements and strategy of the judging have remained constant.

Papers are judged "blind," meaning that the identity of the authors and institution of a paper are unknown to the judges. Each paper bears a code number assigned by the COMAP office that—until the code is broken after the judging is complete—is known only to the authors themselves and the staff member at COMAP who keeps the contest database.

Judging of the MCM papers takes place in two stages, each occupying a weekend: "triage," or preliminary, judging; and final judging. The objective is to classify the solution papers as Successful Participant, Honorable Mention, Meritorious, and Outstanding. The percentages of papers in these classes are approximately 58%, 25%, 15%, and 2%, respectively.

Triage

The triage stage involves one judge for each 25 or so papers, plus a Head Judge. Each paper is scored independently by two judges. They spend ten minutes with each paper, attaching particular importance (about 40%, as determined by the Triage Director and Head Judges) to the paper's summary. If the assessments of the two judges are too far apart, they confer to resolve the difference. If neither judge is willing to change, a third judge breaks the deadlock.

The two scores for a paper are added. The Head Triage Judge, in consultation with the Contest Director, chooses a cut score for the total. About 60% of the papers pass the triage and are sent on to a new set of judges for final grading. The papers culled, if not disqualified for lack of any substance, are awarded Successful Participation.

Final Judging

Like triage, final grading has two sets of judges, one for the "A" (continuous) problem and one for the "B" (discrete) problem, including a judge from the triage team, a Head Judge, and an Associate Director. There is one judge for each 8 or so papers reaching this stage. A standard part of the judging procedure is a preliminary training session in which the judging teams are briefed by the Triage Judge on partial credit, solution methods to expect, etc.

A first "screening" reading takes about 15 minutes per paper. The screening and the previous triage assessments are combined. There are then three more readings, taking about 20, 25, and 30 minutes, respectively. At each stage that it reaches, a paper is read by a different judge. About 30–35% of the remaining papers are culled at each reading, yielding the classification of culled papers as Successful Participation, Honorable Mention, and Meritorious.

For each of the two problems, all of its judges read and discuss the small number of papers that pass all of the readings and cullings. The discussion can become quite heated. A paper is classified as Outstanding only by consensus of each judging team and the Director and Associate Director; papers not classified as Outstanding are classified with the other Meritorious papers. The ORSA and SIAM judges choose their organizations' winners from among the Outstanding papers.

After all judging is complete and judges have double-checked all their results, the Contest Director opens a sealed envelope from COMAP that reveals which teams correspond to which code numbers.

The Voice of Experience

We collect here some tips, suggestions, and strategies contributed by team advisors over the years.

- In late fall, make announcements in upper-level mathematics courses inviting students to participate in the contest. Ask professors and TAs to recommend their most promising students. Emphasize the potential benefits to the participants and the need for a team to include a variety of skills. If there are enough students interested (some institutions have 30 or more), use resumés and interviews as an aid in selection. Look not only for specific skills but also for willingness to work hard and cooperatively in a team.

- Choose team members who have a strong background in writing. The best combination might be two English majors and a mathematics major with an English minor! Each team needs someone who can program well but also someone who knows what canned programs are available. Double majors make good modelers, not only because they have been exposed to a variety of courses and occupations, but also because engineering and physics courses seem to be taught with more of a modeling approach. Someone has to be able to type and be comfortable with the technical word processor available. Also, each team should have some sort of vehicle, if only to get the paper to the post office; I never considered this until one year when one team's sole transportation was a single bicycle among the three members.

- Assign example problems and meet with each team about twice a week during January. If possible, spend a weekend doing a trial run with a problem chosen by the advisor, perhaps from a previous contest.

- If you are offering a modeling course during the spring term, give students who participate on a team appropriate credit toward one of the course projects.

- Chart in advance the hours of the libraries (public and on-campus), scheduled downtime for central computing facilities and networks, and local post offices (including hours for vending postage and dispatch time for express mail service).

- Have students meet with a librarian and go over resources available. Better yet, as part of training, have the team members do a literature search on a particular topic.

- Encourage the team members to write in the first-person plural ("We fit a fifth-degree polynomial to the data...") instead of a stilted third-person passive ("It should be noted that then it was discovered that the data could be fitted by a fifth-degree polynomial. . . "). Encourage them, too, to organized their exposition in logical order; a rambling chronological diary usually doesn't make for a good paper. In addition to numbering all pages, they should number and caption all figures and tables. Teams may find it helpful to read over Maurer [1991] and Gillman [1987] in advance and keep a copy of each handy during the contest.

- Provide a room for each team's exclusive use (with a key to each member) during the weekend of the contest. This could be a departmental seminar room, or a faculty member's office. Arrange with the building support staff for early special attention to cleaning the room on Tuesday morning.

- Provide for 24-hour access to the teams' rooms, computer facilities, and a photocopier (and either a copier credit card or plenty of change), including making any necessary arrangements with campus security.

- If possible, equip each team's room with microcomputers (loaded with available word-processing, modeling, computer algebra system, and statistical software) and a printer. Ideally, a team should be able to function completely independently of any campus computer network, in case the network should fail (or be shut down for backups) over the weekend.

- Equip each team's room with a library of useful modeling resources, including past issues of *The UMAP Journal* and a collection of UMAP Modules. Different teams at the same institution are competing with each other, but it behooves them to share any necessary resources (e.g., relevant library books and modeling resources).

- If possible, provide for all expenses students have during the weekend, including food (e.g., pizza, but the team should get out of the room and eat a decent meal together—or apart—at least once a day), photocopying, any central computing charges, laser printing, and postage.

- Just before the last moment, the team should check the final version of the paper to be sure that all relevant supplements (graphs, figures, tables, program listings) are enclosed.

- After the contest, go over the teams' work for loose ends or omissions and encourage them to complete it. Have them discuss the contest and present their results in a departmental colloquium. Urge them to present their papers in a local or regional student research symposium, in the student papers session of the spring meeting of your regional Section of the Mathematical Association of America, or at the summer MAA Mathfest.

- Have the college's public relations department send out the contest results to local newspapers and to the team members' home-town newspapers.

Acknowledgments

These tips are synthesized from contributions from the following team advisors:

Stavros N. Busenberg (1942–1993), Harvey Mudd College, CA
James R. Choike, Oklahoma State University, OK
G.F.D. Duff, University of Toronto, Toronto, Canada
R.H. Elderkin, Pomona College, CA
Ken Gamon, Central Washington University, WA
J.E. Gayek, Trinity University, TX
Andrew Gleason, Harvard University, MA
Richard Haberman, Southern Methodist University, TX
G.A. Klassen, Calvin College, MI
H.A. Krieger, Harvey Mudd College, CA
J.P. Lambert, University of Alaska Fairbanks, AK
F.H. Mathis, Baylor University, TX
M.M. Meerschaert, Albion College, MI
Thomas D. O'Neil, California Polytechnic State University, CA
Włodek Proskurowski, Reza Iranpour, and Paul Chacon,
    University of Southern California, CA
Pierre Roger, Université Laval, Québec, Canada
Thomas Rogers, University of Alberta, Alberta, Canada
Anita E. Solow, Grinnell College, IA
Philip D. Straffin, Beloit College, WI
Andrew Vogt, Georgetown University, DC
Brian J. Winkel, Rose-Hulman Institute of Technology, IN

References

Gillman, Leonard. 1987. *Writing Mathematics Well: A Manual for Authors.* Washington, DC: Mathematical Association of America.

Maurer, Stephen B. 1991. Advice for undergraduates on special aspects of writing mathematics. *PRIMUS: Problems, Resources, and Issues in Mathematics Undergraduate Studies* 1 (1): 9–28.

A Junior College in the First MCM

Alice Williams
Dept. of Mathematics
James Madison University
Harrisonburg, VA 22807

The Paper

The beauty of the MCM is that students experience government- or industrial-type problem-solving while writing a paper on a weekend at their own campus.

I include below parts of the paper presented in the first MCM by Jody Baker, Jeanine Scheeden, and Lynda Ferguson of Southern Seminary Junior College. They were responding to the problem titled "Modeling an Animal Population," in which they were instructed to select a fish or mammal and an environment, and then to formulate an optimal policy for harvesting members of the population. This team chose the burros in the Grand Canyon of the U.S.

Start with 200 burros, with one-half being female (100 female). Their ages range from 1 to 8 years. Those one year old and under are nonbreedable. One-half of the eight-year-olds are nonbreedable. The result is a breeding herd of 87 females.

Each female has an average of three offspring per lifetime. This was deduced by taking into consideration that the burros would not breed until two years of age; therefore, there would not be a foal until the third year. Also taken into account is the fact that some may be aborted, or stillborn, or the burro may not have conceived in a particular year. If each burro lives approximately eight years, then every year one-eighth are dying. This is 25 burros, half of which would be female.

As their work progressed, the team discovered the differential equations necessary for the rate-of-growth problem [Marcus-Roberts 1983, 46]:

$$\frac{dp(t)}{dt} = ap(t),$$

where a is the rate of increase of the herd. Taking the logarithm of each side, they got

$$p(t) = p_0 e^{a(t-t_0)}.$$

They modeled and then computed their solution on the 1,000 existing animals, considering death rate and birth rate with the parameters of population stability in a deteriorating vegetation situation. This leads to the conclusion that to maintain the herd at 1,000, they must have

$$p(t) = 1000e^{a(0.25 \times 1.0)} = 1295.$$

Thus, 295 burros per year must be harvested for adoption or other park uses.

The contest accomplished for this team exactly what it was designed to do. These students were starting their second semester of calculus, and they extended themselves by studying material ahead in their text [Thomas and Finney 1968]. For example, they learned how to make use of partial fractions to integrate a rational expression. They researched and understood the mathematics that they used from other sources. They learned what the modeling process can accomplish, and they had the experience of a team effort.

The Team

Dr. Ben Fusaro, the founding director of MCM, says that "It is not the three best that make the team, but the best three." The team needs the skills of drafting, typing, and computer programming, plus logic and mathematical maturity and the ability to do research and presentation with style.

There are several ways to choose the team. One is to look for students in the highest-level courses and choose the best combination for the team, making it an honor to be chosen. Another is to announce the contest and invite volunteers. One can also recruit students who are computer-oriented and who appreciate mathematical applications, or choose students who work well together and have productive results.

The team must be trained. Generally, students have not learned how to write a mathematical research paper. They can develop this skill by reading published papers, including published Outstanding papers from previous contests, or by writing papers and having them evaluated. Since at most two teams can enter from any department, it might be good to have a junior team and a senior team. This would allow for experienced team members to move up from the previous year's team.

The Advisor

The advisor is analogous to a coach. He or she must recruit and choose the students, arrange for team registration for the contest, secure a room and appropriate computing equipment, and do pre- and post-competition press releases. The advisor must plan a time schedule with the team, practice, and then give support and encouragement during the contest.

The rewards of this coaching experience are many:

- You will have an intellectual experience with a group of students outside of the classroom.

- You will be respected not only by the team members but also by the academic community.

- You will be helping students learn the art of mathematical modeling.

- The knowledge that you gain of industrial problems and modeling solutions will lead you to better classroom teaching.

- You will be giving your students a first-class ticket into an interview with any employer or applied science department.

On a two-year campus, as well as at four-year schools, there are students who have superior qualities that fit the needs of the team. It is the advisor's job to find those students and help them take advantage of this fantastic opportunity.

References

Marcus-Roberts, Helen. 1983. A comparison of some deterministic and stochastic models of population growth. 1985. Chapter 3 in *Life Science Models*, edited by Helen Marcus-Roberts and Maynard Thompson. Vol. 4 of *Modules in Applied Mathematics*, edited by William F. Lucas. New York: Springer-Verlag.

Thomas, George, and Ross Finney. 1968. *Calculus and Analytic Geometry*. 5th ed. Reading, MA: Addison-Wesley.

About the Author

Alice Williams has taught at high-school, junior-college, and university levels. She has researched at University College, Oxford University, in the area of mathematical modeling. She lives with her husband in Lexington, VA.

Experiencing the MCM at Northcentral Technical College

Robert J. Henning
Mathematics Dept.
Northcentral Technical College
1000 W. Campus Dr.
Wausau, WI 54401

Introduction

I became interested in the Mathematical Contest in Modeling (MCM) in the fall of 1985, after receiving the announcement from Tom Kerkes, Dean of General Education. My first reaction was to discard it, as we do not have a mathematics degree-granting program at Northcentral Technical College.

As luck would have it, David Andrews (a fellow mathematics instructor at NTC) and I were presenting a program at the American Mathematical Association for Two-Year Colleges convention in Memphis that fall. While at the convention, I attended the meeting at which Ben Fusaro and a team advisor discussed the 1985 MCM. As I listened to their presentation, my interest level in the contest, as a true learning experience for the student teams, grew. After the presentation, I asked Dr. Fusaro about the possibility of a two-year post-secondary school sponsoring a team. At NTC, mathematics is considered technical support and is geared toward the majors in which the students are enrolled. I like the concept of teams working together to solve a real-world open-ended problem. Dr. Fusaro clinched it for me when he said, "My goal is for the students to learn to work in teams, using as many material resources as possible, and write their suggested solutions. What is important is not the level of mathematics used but that the students participate in this type of problem solving." It was with this vision that I went back to Wausau and started assembling the first team to represent NTC.

Recruiting Students

When NTC first started sponsoring teams, I personally recruited students from my mathematics classes who appeared to have good mathematics skills and liked solving problems. I would check with the students' major instructors to let them know the kind of students that I needed and to get the instructors' support to encourage the students to participate. Now it is more of a joint effort among Frank Fernandes, Rita Keilholtz, the student life manager, and myself. We put articles in the college newspaper and the weekly newsletter *College Beat*. Rita puts me on the agenda for a Student Governing Board meeting in early fall. I explain what the contest is about and why we encourage students to participate, as well as thank them for their support of the contest. Usually, I will get one or two students who express an interest in participating, and thus begins the list of participants.

Previous students who have participated have expressed great satisfaction with the experience. They really enjoyed the challenge and the learning experience that goes along with the contest. One student participant came back in the fall and told me that the tech reporting course (a required communications class for all associate degree programs) that he took during the summer following his first experience in the contest was more meaningful to him because the experience of the contest. The experience of the contest reinforced the topics covered in tech reporting and made it easier for him to understand the need for good technical reports. He participated for a second year and returned to help supervise the contest the third year.

One year, I assembled a team of three female students. They were talked out of competing by another instructor, and I had to work hard to get them back into the contest. At the banquet to celebrate the participation, one of them got up and stated that it was indeed a true learning experience; and all three were glad that they had participated.

Building Support

To get other faculty support and interest in the contest, I talk to those whom I know have a desire to see the students learn more than just classroom learning. Once I explain the merits of the contest—team participation, and mathematics and writing skills—I find them willing to support the experience. I talk about the contest at departmental meetings and at general education divisional meetings, whenever I can get the contest on the agenda. As new faculty come to our campus and into the General Education Division, I share the contest information with them and seek their assistance. The first few years, I was doing everything myself: getting teams together, arranging for the use of the facilities, shopping for the food, and seeing that the papers were mailed on time. Now it has become more of a faculty team putting things together and assisting in all phases of the contest.

We first sought financial support from business and industry in 1993–94. We requested a sponsorship of $250 per team, for which the sponsor gets an invitation to the banquet, a plaque, and their firm's logo printed on a T-shirt given to each participant, advisor, and sponsor. The effort raised

$1,300. The banquet is hosted by the college president, a tradition since the first contest. Each student also receives a college paperweight engraved with their name and year of participation. At the banquet, the students share their experiences with all present and receive a printed booklet of the year's NTC team entries. Also, a copy of the booklet is sent to the library, to become a permanent part of its collection.

At the 1994 banquet, we had a good turnout of sponsors. They felt that the contest was very interesting and a rewarding experience for the students. The banquet gives a time for the sponsors to meet with the college president, faculty, and students in an informal setting. I believe it gives the employer sponsors a closer tie to the college and helps them to understand what our students are capable of doing, in and out of the classroom.

Training

The training and precontest information is done between January and the contest date. We have four meetings, with all the students present. The first meeting introduces the students to teamwork and what mathematical modeling is about. The second meeting, done by the Communications Dept., stresses the important elements of writing a technical report and the brief summary needed. During the third and fourth meetings, we train the students on the computers and the various programs that are available to them, as well as find out what they already know about computers and calculators. In 1994, some of the students used the TI–85 graphing calculator to run simple programs, testing them before setting them up on the computer.

In 1995, we offered a one-credit-hour class in mathematical modeling (and also gave a second credit-hour to those who participated in the contest). The course develops the modeling mindset for the student, as well as showing the importance of teamwork. We have 18 hours available to do some modeling problems and develop understanding of the modeling process. It is mostly hands-on, with very little teacher-led lecture or demonstration.

Conclusion

I would like to relate a story from one of the participants in the very first contest. The reason that I recruited this student was that she had a calculus background and in less than nine weeks had finished the Tech Math 1 course and the first half of Tech Math 2, both with a 98% average. I also found out in talking with her that she loved problem solving.

At the banquet, she asked how knowledge of problem solving could be made known to potential employers when it was not shown on her transcript. I told her to include it in her resumé.

In the spring after graduation, she stopped by and stated that she had been hired by a company in Minnesota and that she felt that she got the job partly due to her experience in the modeling contest. During the interview, the personnel manager had seen on her resumé mention of "MCM 86" and asked her what it was about. After her explanation, he stated that he loved the idea and concept of the contest. This one comment has kept me working on keeping the contest going at NTC for our students.

About the Author

Bob Henning was born in Milwaukee. When he was in the fourth grade, his family moved to Grafton, Wisconsin, where he attended a one-room schoolhouse. In the eighth grade, he had a teacher whose hobby was mathematics.

"He gave a problem of the week to challenge us. Through these problems and his tutoring, my love for mathematics began and grew. In high school, I took all the mathematics I could, as well as industrial arts. I went to college to become an industrial-arts teacher; in my freshman year, another mathematics instructor influenced me to minor in mathematics."

Bob has a B.S. in industrial arts education with a minor in mathematics and an M.S. in vocational education from the University of Wisconsin–Stout. Over the last 30 years, he has taught mathematics to vocational students enrolled in such fields as machine tools, agricultural mechanics, welding, and prep algebra. Northcentral Technical College has awarded him the title of Master Teacher.

"My richest experience was teaching blueprint-reading and mathematics to apprentices from all the various trades. Currently, I am working on a second M.S. degree, in mathematics education, from the University of Wisconsin–Oshkosh. It was the contest and Dr. Fusaro that motivated me to try for it; the contest put everything into perspective for me in terms of problem solving, communications, and applied mathematics, as well as the needs of industry in team problem-solving across the disciplines. My rich experiences from industry and my need to teach students to use mathematics as a tool keep me working and studying in this arena. A personal thank-you to COMAP for ensuring that the contest keeps going!"

Modeling as a Precursor and Beneficiary of Mathematics Reform

Henry J. Ricardo
Dept. of Mathematics
Medgar Evers College (CUNY)
1150 Carroll St.
Brooklyn, NY 11225–2298
herme@cunyvm.cuny.edu

Introduction

The current undergraduate mathematics reform movement owes a great deal to the increased visibility and vitality of undergraduate mathematical modeling in recent years. I've reached this conclusion through working with the materials produced by various reform mathematics projects and by gaining greater familiarity with the MCM. The first MCM was held in 1985, so that both the contest and *The UMAP Journal* itself antedate the famous (or infamous) Tulane conference of 1986, generally regarded as the cradle of the calculus reform movement.

In turn, this reformation should foster greater interest in modeling and elicit even better performances in the MCM.

Evidence

Certainly, the similarity of language and tone between reform literature and modeling literature suggests that the paradigm, explicitly or implicitly, for much of the current reform movement in mathematics is mathematical modeling. For example, compare the following statements:

> Most of the problems in the book are open-ended. This means that there is more than one correct approach and more than one correct solution. Sometimes, solving a problem relies on common sense ideas that are not stated in the problem explicitly but which you know from everyday life.
>
> — Preface to Hughes-Hallett et al. [1994]

Problems will tend to be open-ended and are unlikely to have a unique solution.

— MCM brochure

Indeed, the problems in the Harvard Consortium calculus text [Hughes-Hallett et al. 1994] that I have been using for the past four semesters are both stimulating and frustrating to students because many do not follow a pattern established in the chapter examples. These problems require *thought*. Although sections on differentiation formulas or integration techniques have a fair number of drill problems, there are always thought-provoking exercises that require numerical, graphical, or verbal solutions rather than algebraic manipulations—exercises for which there may be no "right" solution.

Even at a time when textbooks are using a much greater variety of interesting, "real-life" problems, what often is presented in a traditional exercise is the *end* result of mathematical modeling—in calculus, an analytical expression to be differentiated or integrated—rather than any practice in formulating a model.

The current reform movement, at whatever level [Boyce 1994; Callahan et al. 1993; Gordon et al. 1995], is rooted in the realities of mathematical applications. It recognizes that functions don't drop at a student's feet begging to be differentiated or integrated. There is experimentation. Students collect data, plot points, and perhaps fit curves. In other words, a practical situation is *modeled*. Chronologically, the explicit analytical representation of a function may be the last representation of a fundamental relationship between or among variables. In a modern differential equations course, for example, there is less emphasis on closed-form solutions than on numerical approximations and the *analysis* of solutions (see, for instance, [Boyce 1994]).

There are more culture shocks awaiting students accustomed to manipulation of symbols:

The keys to getting the most out of these problems are thinking, discussing, and writing .... Finally, write up your conclusions in complete English sentences that convey your understanding as clearly as you know how. With practice, you will discover that discussing and writing promote clear thinking and thus help you develop a better understanding of the material that you are studying.

—"Suggestions to the Student" from Fraga [1993]

Attention must be paid to the clarity, analysis, and design in attempting a solution. The narrative part of the solution papers must be typed and in English.

— MCM brochure

Let's take a closer look at some of the guiding principles of current mathematics reform.

In general, most reform calculus and precalculus courses attempt to give equal weight to numerical, graphical, and algebraic approaches, thereby diminishing the traditional emphasis on the analytical aspect. (This is called the "Rule of Three" in Hughes-Hallett et al. [1994].) The use of technology such as graphing calculators and computer algebra systems can be considered either a motivating factor for or a consequence of the emphasis on numerical and graphical descriptions of functions and problem situations in a reform course.

Another fairly common pedagogical thread running through reform materials is the motivation of formal definitions and procedures by the investigation of practical problems (the "Way of Archimedes" in Hughes-Hallett et al. [1994]). In other words, theory is shown to develop naturally from practical problems. For example, one of the reform projects develops calculus concepts from the perspective of differential equations and starts the course with a model of the spread of a contagious disease [Callahan et al. 1993]. Of course, historically, this has often been the case. Those committed to reform claim that this type of course develops thinking skills by fostering experimentation and discovery. Rote learning gives way to mathematics as a laboratory science. Teaching techniques described as "cooperative learning" or "collaborative learning" usually are seen as important in any reform course. Such a course also develops communication skills by encouraging reading, writing, presenting, and group work.

In practice, the Rule of Three is often augmented by a fourth principle advocating the oral and/or written analysis of concepts (see Sterrett [1992] for valuable insights). The MCM announcement letter from the Contest Director to department heads states that the MCM is "designed to stimulate and improve problem-solving and writing skills in a team setting." Indeed, this emphasis on team work is one feature that distinguishes the MCM from its prestigious forebear, the William Lowell Putnam Mathematical Competition.

Conclusion

In many ways, "lean and lively" reform mathematics de-emphasizes conventional content while actually teaching the basic skills needed for mathematical modeling:

- deriving a mathematical formulation of a practical problem,

- analyzing ("solving") the resulting mathematical problem(s), and then

- interpreting the results in terms of the original physical situation.

Consequently, a broader range of students will benefit from this basic training in the mathematical sciences, while some of those captivated by doing mathematics in this way will be able to pursue more rigorous theoretical work later.

Although relatively few high achievers on recent MCMs may have benefitted from exposure to reform mathematics, I am confident that as the reform movement spreads, the level of achievement will rise proportionately.

References

Boyce, William E. 1994. New directions in elementary differential equations. *College Mathematics Journal* 25 (5): 364–371.

Callahan, James, David Cox, et al. 1993. *Calculus in Context*. New York: W.H. Freeman.

Fraga, Robert, ed. 1993. *Calculus Problems for a New Century*. Vol. 2 of *Resources for Calculus*, edited by A. Wayne Roberts. Washington, DC: Mathematical Association of America.

Gordon, Sheldon P., B.A. Fusaro, et al. 1995. *Functioning in the Real World: A PreCalculus Experience*. Class Test Edition. Reading, MA: Addison-Wesley.

Hughes-Hallett, Deborah, A.M. Gleason, et al. 1994. *Calculus*. New York: Wiley.

Sterrett, Andrew, ed. 1992. *Using Writing to Teach Mathematics*. Washington, DC: Mathematical Association of America.

About the Author

Henry Ricardo graduated with honors in mathematics from Fordham College and received a Ph.D. from Yeshiva University in 1972. After 12 years teaching at Manhattan College, he began a 14-year stint with IBM by working on a mathematical model of software-defect discovery and its implementation in APL.

Since returning to academe as Associate Professor of Mathematics at Medgar Evers College of the City University of New York, he has been active in teaching reform mathematics—precalculus, calculus, and differential equations. He has also co-edited a guide to mathematical modeling, with Dr. John Loase of Westchester Community College, under a three-year NSF grant.

Ten Years of MCM: Reflections of a Coach

Thomas O'Neil
Mathematics Dept.
California Polytechnic State University
San Luis Obispo, CA 93407
toneil@oboe.calpoly.edu

Introduction

In the fall of 1984, our department received the announcement of the first Mathematical Competition in Modeling. Since I was the coach of our Putnam Competition teams, the information was thrown into my mailbox. The flyer stated that one of the objectives of the competition was to encourage mathematics departments to establish courses in mathematical modeling.

We then had already had a modeling/problem-solving course for seven years, serving as our capstone course and required of every mathematics major. Whenever we queried our alumni about our program and how well we prepared them for the real world, this course was always a runaway first choice. The emphasis on problem solving and communication skills were the most frequently stated reasons.

How did our modeling class stack up to others? I had the same problem back then that we run into today. Everyone has an innovative idea that they want to use in the classroom: a leaner calculus, fewer drill problems replaced by problems requiring creative thought, handheld calculators, and a computer at every desk with Mathematica (or maybe Maple V is better). But how do you evaluate any of these approaches? The MCM was just what I had been looking for. Our teams did very well the first year and have done well ever since; I had the information I needed.

Our original modeling course emphasized individual problem solving, with most of the problems possessing unique solutions. Since the MCM is geared to a team format and open-ended problems, I created a new course in which the students are put in teams and given problems of this type. I generally give them two weeks to get their solution papers back to me. Then I stir up the team memberships, give them another problem, and get to work criticizing the writing of their submitted papers.

How to Pick a Team

I firmly believe that the best way to select a team is to find three English majors, one of whom is working on a math minor. The writing is so important! In this respect, the contest really reflects reality. I think that in our classes in mathematics, we don't emphasize the importance of communication skills nearly enough. We require our students to take a lot of English and speech classes, but how often do they have the opportunity to use in a mathematical setting what they have learned in these courses?

Putting the students on teams in the modeling course lets me detect personality conflicts. Four days can be a long time to work under stress. Success on the MCM says as much about the personality of the contestants as their mathematical ability. I don't make teams out of friends. Friends look too much alike, and I have found that diversity is a key ingredient for a good team.

Each team needs one person who has what I refer to as a "terrier personality": someone who is jumpy, coming up with a new idea or approach every thirty seconds or so. You know the kind of student that I am talking about, the kind of person usually does well on the Putnam Competition. You can't have two of them on the same team, at least not for four days.

Then there has to be a member who can lend a degree of stability to the team—someone who can keep the terrier reined in, with a leash if necessary. This kind of student is easy to find. These are the ones who are in class every day with their homework done and know what they are going to be having for dinner two weeks from next Tuesday.

The third person has to have a personality that is a mix of the other two, and I generally choose a student who has already had most of the upper-division courses required of our mathematics majors.

One of the team members has to be a programmer, and one has to know lots about statistics. They all have to be sharp, they all have to be willing to work until they drop, and—once again—*they all have to know how to write.*

Some of our best results have come with students who are either double-majoring or taking a minor outside of mathematics. The more that the team members have seen, the better the odds of someone having some idea of what to do with the MCM problems.

One of the most difficult jobs that I have is to tell all but six class members that they are not being chosen to participate in the contest. These students work hard for a quarter in hopes of being one of the chosen few; and the disappointment on their faces is heart-wrenching. At least in the Putnam Competition everyone gets to play, whether they are on the team or not.

Support and Recognition

The easiest way to recruit new talent each year is to give a lot of recognition to the members of the current teams. Our display case, which contains several MCM plaques and records of MCM and Putnam results, is located in a high-traffic area. We have a department meeting each spring at which we present student awards, and we give the MCM team members their certificates on this awards day. We also take team pictures and put them in our annual alumni newsletter, along with an article about the contest and the results. Additionally, we have each team present its solution paper at a faculty colloquium and invite the rest of the students.

Our students get a lot of support from not only the faculty but the staff as well. We have a Division I sports program at Cal Poly; but due to financial problems in the academic area, our library closes at 5 P.M. on Friday and does not open again until Saturday at noon. Other than their own talents, the most important tool the students have for this contest is the library; and ours is closed at the most crucial time of the event. But our library staff has bent over backwards to help us. Every year, they have had someone either stay late on Friday night or open early on Saturday for the team members. For a few years, the library even remained closed on Saturdays; but the library staff came through every time, sending someone in to open up for the participants. One of the reasons that library staff are so cooperative is that there is a plaque, presented by the Mathematics Dept., hanging in the library thanking them for their continued support of our MCM teams. The best way to get someone to do something for you twice is to thank them the first time.

In 1985, no member of the Mathematics Department faculty had a computer in his or her office. The students used their own Macintoshes to do the writing and what little computing was needed on the problems that year. The next year, the university saw its way clear to buy two computers. On the Thursday before the contest, I simply went to the people with the machines and told them we needed them for a few days. No problem. Nowadays, each office is equipped with some sort of computing equipment and the faculty members use their machines every day: receiving and sending e-mail, looking up student and class records, writing exams, and even doing a little mathematics. It has become difficult for most of us to do without our computer for four days. However, the majority of the faculty do feel that this event is important enough to let teams borrow whatever is needed.

The Effect on Team Members

Over the years, many of our students who have participated in the MCM with success have been those whose academic record would not predict this success. For some reason, even though they are bright, many of these students just don't seem to apply themselves enough to receive high grades. Yet some of these same students, when confronted with a problem that appears to have come from real life, will work their hearts out for four solid days. One of the most amazing things that I have seen over the years has been the rise in confidence that these students have when they receive the results of the judging of their work. Almost every one has decided to go on to graduate school, and they have done so successfully. With their academic records, these same students would probably have had trouble finding jobs as mathematicians; but with the confidence they gained through the contest, they have attained goals that they would not have dreamed possible before their participation. This fact alone makes it worthwhile for our school to field teams each year.

As I was just writing the above, I was interrupted with a phone call from a former student and MCM veteran. He has been working for a consulting firm for several years now and called to inquire about a recent graduate who had applied for a job. When I mentioned the fact that the young lady had received an A in my modeling class and that her team had received a Meritorious MCM award last year, he immediately decided to hire her.

A Few Misgivings . . .

There have been a few things that have disturbed me about the contest over the years.

- My primary gripe over the years was that, until recently, the cutoff time for the contest was dictated by a Monday postmark. Since our local post office closes at 5:30, I always had to impose a 5:00 deadline on our students. However, if we were located in a larger city, we could still get a Monday postmark at midnight. Ben Fusaro hated to see me approaching him at mathematics meetings, because he knew that before the conversation was over I would broach the matter of a common cutoff time. His response (always in jest) would be "Your teams don't need the extra time." Once he asked how far it is to the nearest post office with a midnight postmark. When I conjectured that it might be Santa Barbara, one hundred miles south of us, he said, "That would be twenty minutes in California time. Just drive down and mail from there." My message must have gotten through, because, under a change in the rules two years ago, all teams must now quit by 5:00 local time. Now I feel better, and Ben no longer has to try to hide from me at meetings.

  This might be a good spot to say how appreciative I am of the work that Ben has done with the contest. Even now that he is no longer its director, he continues to be involved in everything from finding problems to judging. It is still his baby.

- A problem in the 1988 contest asked how to load as many boxes as possible, from a collection of boxes of various weights and sizes on a railroad siding, onto two flat cars. I have often wondered if the boxes that did not fit onto those two flat cars are still sitting on the siding. Why not make the problem realistic and ask for the minimum number of flat cars necessary to haul away all of the boxes?

- And then there was the 1992 contest with the air-traffic-control-radar problem. It was full of unrealistic information. One could smell a rat immediately upon looking at the figures that were given. One of these was the end view of an ellipsoid of revolution, looking down the axis of revolution. The figure was an ellipse! The students not only had to fight their way through the misinformation but also had to worry about whether they should be solving the problem as stated or to use more realistic data and solve their own improved problem. They ended up doing both, but not until after a lot of fighting among themselves. The Judge's Commentary published in *The UMAP Journal* includes a statement to the effect that this problem is realistic in that it is ill-posed, as are most problems that consultants are given. When I was in the industrial game, an engineer didn't leave my office after giving me a problem until we had discussed the problem and I had his phone number. If there were any questions, I called. These students do not have that luxury.

... **but Rich Rewards**

The personal rewards that I have received from being involved in the MCM all of these years far outweigh any of the aforementioned complaints. I have met a lot of people and have established many good friendships. When one of our students on an Outstanding team was interviewed by the campus newspaper, he attributed the team's success to a faculty whose only reward is the success of their students. A statement like that can't help but make you feel good. I would like to encourage anyone who likes to work with students to get involved. It is difficult to get old being around all of the enthusiasm of young people.

About the Author

Thomas O'Neil received A.B. and M.S. degrees from San Diego State College and a Ph.D. in mathematics from the University of Wyoming. He is currently Professor of Mathematics at California Polytechnic State University in San Luis Obispo, where he has taught since 1973. His current interests include problem solving, modeling, applications of mathematics, and the pedagogical issues associated with computing in the mathematical classroom. His background includes ten years of experience as technician and engineer in electronic research and development, with the U.S. Navy, General Dynamics/Astronautics, and The Boeing Company.

The MCM Experience at the University of Alaska Fairbanks

J.P. Lambert
Dept. of Mathematical Sciences
University of Alaska Fairbanks
Fairbanks, AK 99775–6660
ffjpl@aurora.alaska.edu

Introduction

The University of Alaska Fairbanks has had a vibrant relationship with the MCM. From interpolating hydrographic data in 1986 to designing a faculty compensation plan in 1995, eighteen UAF teams have faced the MCM weekend and have without exception thrilled to the challenge. A special highlight came in 1990, when a UAF team's paper on snow-plow routing was judged Outstanding. Another Outstanding team followed the next year (the first all-woman Outstanding MCM team, with all three students, it happens, graduates of the same Fairbanks public high school); there have been two more since then, in 1993 and 1995.

The MCM has indeed been good for UAF. Ben Fusaro's inspiration more than a decade ago has proved a superb educational experience for our students. They have learned to work and think and produce a substantial report in close collaboration with others under a strict time constraint; they have learned new mathematics and new applications of mathematics. As faculty advisor to UAF teams each year except 1991—when I was on sabbatical in Ireland and present at the start of now vigorous MCM traditions at Trinity College Dublin and University College Galway—I will try to share some thoughts about what happens here in Fairbanks.

Finding and Training the Teams

Team selection and composition are of course important. We seek balance and breadth of expertise, and compatability. Computer and writing skills are essential; statistical knowledge is valued if available. Students have come to MCM teams from a variety of majors: physics, geology, engineering, chemistry, economics, statistics, biology, computer science, and mathematics. A number have had double majors. Nearly all have taken at least one upper-division mathematics course, especially from among mathematical modeling (which most have taken), linear algebra, numerical analysis, discrete structures, and advanced calculus. Often prospective participants have first become known to us in such courses. Also, an announcement about the MCM is posted on a bulletin board, inviting inquiries; I am always interested in discussing the contest with students enticed in this way. So far, there has been no dearth of eager talented students available, many of whom were attracted to UAF in the first place by its strengths in the sciences and by the uniqueness of Alaska (including some from abroad: Russia, Sri Lanka, Greenland). Most of our MCM participants have in fact been from within the state.

We try to identify prospective teams (usually two) early on, by the first of December in the semester preceding the contest, and to establish a weekly or biweekly meeting routine. It is crucial to get together with the students regularly, especially in the period after semester break leading up to the competition. Students are given copies of past MCM problems, and solution papers are available to them; they are encouraged to identify with each other as teams and to meet separately to discuss problems, strategies, and division of responsibility. Ideally, it would be nice to set up a dry run of the overall MCM weekend; but no one has time for that sort of commitment. At our collective meetings each week, we talk about the nature of applied mathematics and about problem solving; but much of the time, we talk also about computer facilities and software, library and departmental facilities, MCM rules, and the importance of a well-written summary. Sometimes we have guest speakers. A department head once proposed that MCM preparation and participation be formalized as a course with credits, an idea that (despite some appeal) was rejected. One participant warned that this would eat at the special nature of the MCM and lead to a class-like situation, with "professors giving stuffy lectures, and grades," as he indelicately put it to this professor.

The Long-Awaited Weekend . . .

The weekend of the MCM is always exciting. There is a charged air of anticipation when the envelope is finally opened just past midnight on Friday and the two problems compete for attention. By this point, various details have been attended to. Each team has been issued a key to the Chapman building (which houses the Dept. of Mathematical Sciences), to the departmental office, and to my office. Accounts have been set up on computers; campus security has been alerted to the likelihood of goings-on at all hours.

Over the years, a strong tradition of faculty support for the MCM effort has evolved at UAF: Teams are given the run of much of the building; and as the weekend unfolds, pizzas, sandwiches, fruit, Chinese food, cookies, juice, and soft drinks appear in the faculty lounge at regular intervals. So sustained, the teams forge on through setbacks and disagreement, breakthroughs and discovery, sleep becoming a diminishing commodity as the weekend advances. On Monday, sometimes perilously close to deadline, a solution paper always materializes. Two copies are produced for COMAP, one for each team member, and one for me; and the MCM is concluded for another year.

... and After

But not really concluded. The annual MCM undertaking at UAF does not quite end with the contest itself on a February weekend. There is a special event in late April, tied in when feasible with Mathematics Awareness Week, at which teams are honored and invited to give short presentations of their solutions. Student attendance is encouraged and pizza is provided. More than one future participant's interest in the MCM was first sparked by this event, through seeing what other students could accomplish and being witness to their enthusiasm.

About the Author

Pat Lambert was educated at Xavier High School in Cincinnati, the University of Cincinnati (B.S.), the University of New Mexico (M.A.), and the Claremont Graduate School (Ph.D.). He has taught at a Nigerian secondary school, at Ahmadu Bello University in Nigeria, and at University College Cork in Ireland, and has worked at a government research lab in Maryland. Since 1982, he has been at the University of Alaska Fairbanks, where he is Professor of Mathematics, and has had visiting positions at Trinity College Dublin and University College Galway in Ireland. His interests include quasi-Monte Carlo methods, voting theory, and mathematical modeling.

1985: The Strategic Reserve Problem

Cobalt, which is not produced in the U.S., is essential to a number of industries. (Defense accounted for 17% of the cobalt production in 1979.) Most cobalt comes from central Africa, a politically unstable region. The Strategic and Critical Materials Stockpiling Act of 1946 requires a cobalt reserve that will carry the U.S. through a three-year war. The government built up a cobalt stockpile in the 1950s, sold most of it in the early 1970s, and then decided to build it up again in the late 1970s, with a stockpile goal of 85.4 million pounds. About half of this stockpile been acquired by 1982.

Build a mathematical model for managing a stockpile of the strategic metal cobalt. You will need to consider such questions as:

• How big should the stockpile be?

• At what rate should it be acquired?

• What is a reasonable price to pay for the metal?

You will also want to consider such questions as:

• At what point would the stockpile be drawn down?

• At what rate should it be drawn down?

• What is a reasonable price at which to sell the metal?

• How should sold metal be allocated?

Below we give more information on the sources, cost, demand, and recycling aspects of cobalt.

Useful Information on Cobalt

The government has projected a need of 25 million pounds of cobalt in 1985.

The U.S. has about 100 million pounds of proven cobalt deposits. Production becomes economically feasible when the price reaches $22/lb (as occurred in 1981). It takes four years to get operations rolling, and then six million pounds per year can be produced.

In 1980, 1.2 million pounds of cobalt were recycled, 7% of total consumption.

Please see **Figures 1–3**, whose source is *Mineral Facts and Problems*, U.S. Bureau of Mines (Washington, DC: Government Printing Office, 1980).

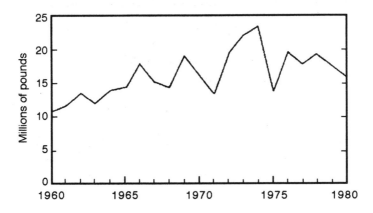

Figure 1. U.S. primary demand for cobalt, 1960–1980.

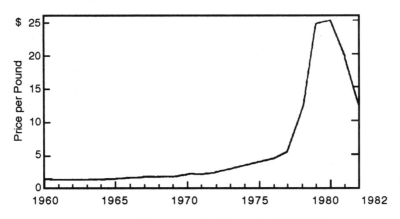

Figure 2. Cobalt prices in the U.S. market, 1960–1982.

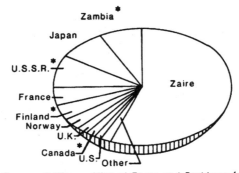

Source: U.S. Bureau of Mines , Mineral Facts and Problems (1980)

Figure 3. Producers of refined metal and/or oxide, 1979. An asterisk denotes a country with domestic production.

Comments by the Contest Director

The problem was contributed by Fred Roberts (Dept. of Mathematics, Rutgers University, New Brunswick, NJ).

Managing a Strategic Reserve

David Cole
Loren Haarsma
Jack Snoevink
Calvin College
Grand Rapids, MI 49546

Advisor: G.A. Klassen

Abstract

We develop a method of managing strategic reserves, in this case the U.S. strategic cobalt reserve. We present a rationale for the stockpiling of cobalt, followed by a method for bringing the stockpile amount from any level to a desired goal, in this case an amount determined by the Federal Emergency Management Agency (FEMA) to last through three years of conventional warfare. The method involves balancing the expected social benefit, based on an estimate of the probability of war or a major supply disturbance occurring before the goal is realized, and the social cost, determined from the additional amount that U.S. cobalt consumers have to pay due to the increase in world demand brought about by stockpiling. We then discuss management of the filled stockpile, with the idea of using the stockpile to assure stability in the world price of cobalt and to defray maintenance costs by market speculation. We consider the conditions under which the stockpile should be drawn down, with the proper rate and total amount of released stockpile material determined for two cases: a major supply disruption, and actual warfare. Finally, we generalize the method to cover other strategic stockpiles.

Introduction

In 1946, the Federal Government decided to establish stockpiles of certain minerals that were deemed important for national defense and that were, for the most part, imported from other countries. The metal *cobalt* is one material that has been stockpiled since the program's inception. It is used in many important industries and is no longer mined or produced domestically. We will investigate some of the considerations in creating and maintaining a stockpile of the strategic metal cobalt as an example of the general problems of stockpile management.

We begin by showing the current necessity of a cobalt stockpile, due to the many uses of cobalt and the current vulnerability of its supply. We continue by developing a model for filling the stockpile and managing the stockpile once it has reached its capacity. We then consider the criteria and methods for directing the draw-down of the stockpile during crisis situations. Finally, we discuss extensions of these methods to other materials.

The Problem of Vulnerability

Cobalt is used in many high-technology applications: "superalloys" for aviation technology, cementing compounds for high-tensile-strength tools, magnets in electrical equipment, dryers and pigments in paints, and so forth. Many of its applications have strategic value in a possible conflict; in 1980, defense used 2.9 million pounds or 17% of that year's U.S. consumption, and during the Vietnam War cobalt consumption increased 39%. Unfortunately for these industries and for national defense, there is no domestic production of cobalt [Congress of the U.S., Congressional Budget Office 1982; Sibley 1979; Veinott 1965, Chs. 7, 9].

These facts indicate that the U.S. is vulnerable to disruptions of the supply of cobalt. There are two primary concerns: disruptions caused by problems in supplier countries, and those caused by war.

The first, shortages and price increases due to an upheaval in an unstable source of supply, already took place in the 1970s. In September 1976, FEMA, which oversees the stockpile program, increased the goal for the cobalt stockpile and ceased selling what had been a surplus of the metal. This caused a tight market and some price increases. Then, in 1978, a political uprising in Zaïre brought production in that country to a halt and caused a panic. Prices jumped from $2/lb to $25/lb, with prices of $50/lb on the spot market. The problem in Zaïre was resolved in less than a year, but prices of $25/lb continued into 1981 [U.S. Dept. of the Interior, Bureau of Mines 1983].

The second type of disruption is what the Strategic and Critical Materials Stockpiling Act was intended to mitigate—disruption due to severed supply lines and increased demand during wartime. Currently, according to the provisions of the Act, this is the only legitimate use of the U.S. stockpile; however, we believe that there is also merit to using the stockpile to alleviate the first type of disruption.

Other concerns are

- the possibility of a cobalt cartel,

- the abilities of supplying countries to meet future increases in demand, and

- the almost inevitable price increases as accessible sources are used.

Approaches to Decrease Vulnerability

The creation of a strategic cobalt stockpile is not the only solution that has been suggested for the above problems. In fact, for the long-range problems of availability and price, a stockpile is at best useless and at worst a means of treating symptoms if it were drawn down during the beginning stages of a significant problem. Such long-range problems require long-range solutions.

During and after the Zaïre crisis, there was considerable research into alternative solutions. The U.S. has 100 million pounds of proven cobalt reserves, which could be brought on line to produce six to eight million pounds per year if the mine operators were give a three- to four-year lead time and a guaranteed price of $25/lb for the next ten years. Recycling and less wasteful manufacturing techniques resulted in some savings after 1979 and could result in further savings if the producer prices for cobalt increase. Research to discover and improve materials that can substitute for cobalt but are more readily available, as well as research into ocean-floor mining, may also provide help for the long-range picture [Congress of the U.S., Congressional Budget Office 1982; Sibley 1979].

None of these solutions is available on short notice, however. Domestic reserves, recycling, and substitutes can all be considered stockpiles that require a certain initial expenditure and lag time before they can be "drawn down." Also, there is a cost associated with drawing them down, so that they are worth using only if the price of cobalt risers more than a certain amount for an extended period of time.

Thus, some short-term solution is required for the disruption caused by a war or an unstable supply; a stockpile is well suited to those cases. Currently, the stockpile size goals are set by FEMA. It uses an interdepartmental procedure to calculate the requirements of a wartime industry for strategic materials during a three-year war fought with conventional (non-nuclear) weaponry. It does not make its criteria publicly available, for security reasons; thus, as mathematicians all we can do is accept their findings and work with them. The present stockpile goal is 85.4 million pounds, 45 million of which had been stockpiled by 1983 [U.S. Dept. of the Interior, Bureau of Mines 1983].

Reaching the Stockpile Goal

Perhaps the biggest concern of stockpile management is that of ensuring that the stockpile is of adequate size; in this case, enough cobalt to last through a conventional war as determined by FEMA. At present, the stockpile is too low and needs to be built up. In the future, stockpile requirements may be changed at any time. Stockpile management requires a means of determining the rate at which cobalt should be bought.

The simplest method is to purchase at a constant rate. If A is the amount of cobalt needed and t is the time until the target date (the date determined when the stockpile should have reached its goal), then R, the rate at which cobalt should be bought, is simply $R = A/t$.

Thus, if the current month is February 1985 and the current stockpile is at 51.4 million pounds of cobalt, while the target stockpile is 85.4 million pounds and the target date is December 1991, then the amount of cobalt needed, is $A = 34.0$ million pounds of cobalt with time $t = 83$ months. In this case, the recommended rate of purchase is $R = 34/83$ million $= 410$ thousand pounds of cobalt per month (4.9 million pounds per year). A constant purchasing rate is the simplest, and for a first approximation, the best, since constant purchasing keeps the market as steady as possible and reduces the chances of undesirable fluctuations in price (due to fluctuating demand). Some reasons and models for purchase and sales of cobalt at nonconstant rates are given below.

It should be noted that this model works more generally than just for initial stockpile buildup. Suppose that the target stockpile of cobalt has been reached. If for any reason the required amount of stored metal is increased, then this new target, along with a new target date, gives a new purchasing rate. If, on the other hand, the required reserve amount *decreases*, then A is negative, R is negative, and the formula gives the rate at which cobalt should be *sold*. During stockpile buildup, any time a purchase above or below the constant rate is made, a new rate should be calculated.

Benefits of Accelerated Purchasing

If there is a goal stockpile level, then what is the optimum rate at which to reach that level? We wish to answer that question by looking at a *stochastic programming problem*. We want to minimize, subject to certain constraints, the expected value of the cost of a solution in which certain costs have an associated probability distribution for their actual occurrence. Specifically, there will be costs associated with having a shortfall in the stockpile if a war or disruption should occur, which should be weighted by the probability of such a disruption actually occurring. Only if we *knew* ($p = 1$) that a war would occur would we build up at a very high rate and risk disrupting the market.

We can simplify our problem by grouping possible future events into three cases (more could be used): war disruption (denoted w), supply or price disruption (s), and no disruption (n). Assigning a probability to each of these events (p_w, p_s, and $p_n = 1 - (p_w + p_s)$), our stochastic problem becomes the nonlinear programming problem to minimize $p_w c_w + p_s c_s + p_n c_n$, where the c's represent the costs associated to each event.

Let us consider that question of the social costs of a disruption as a function of the *fraction* of stockpile complete. Thus, if s is the stockpile level and g is the goal, the *cost* ξ is $\xi = \xi(s/g)$. We would expect the following to hold:

- the domain of ξ is $[0, 1]$;

- $\xi(0) = \alpha > 0$, where α is the cost with no stockpile, and $\xi(1) = 0$; and

- $\xi'(0) < -1$, $\xi'(1) = 0$, and $xi'(x) \leq 0$ for $x \in [0, 1]$.

The expectation that the cost with a full stockpile is zero is unrealistic, but we can move the axis to give us this zero for easier computations. That is, we expect the curve to look like **Figure 1**; the closer the stockpile is to the goal, the less expensive a disruption will be.

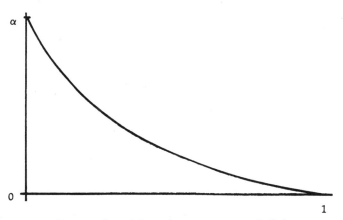

Figure 1. Cost of disruption vs. amount stockpiled.

Integrating the cost over one year, we obtain the expression below for the yearly cost of a buildup in which the stockpile contains level(t) pounds of cobalt at time t:

$$\int_0^1 \xi\left(\frac{\text{level}(t)}{g}\right) dt.$$

For a specific example, let $\xi(x) = \alpha(1 - x)^2$ and consider buying at a constant rate m. Then the level at time t is level$(t) = mt + s_0$, and the cost for one year is

$$\int_0^1 \xi\left(\frac{\text{level}(t)}{g}\right) dt = \int_0^1 \alpha\left(1 - \frac{mt + s_0}{g}\right)^2 dt$$

$$= \frac{\alpha}{g^2} \int_0^1 (g - s_0 - mt)^2 \, dt$$

$$= \frac{\alpha}{g^2} \left((g - s_0)^2 t - m(g - s_0)t^2 + \frac{m^2 t^3}{3} \right) \bigg|_0^1$$

$$= \frac{\alpha}{g^2} \left((g - s_0)^2 - m(g - s_0) + \frac{m^2}{3} \right).$$

In terms of m, this is a parabola with minimum at $m = 3(g - s_0)/2$. From this, it is clear that increasing the rate of cobalt purchase beyond the constant purchase rate will decrease the potential cost of a disruption. In fact, if you consider only this equation, the best plan is to buy all the cobalt that you need right now.

We can also use nonlinear functions for a buying rate and approximate any arbitrary buying function with a piecewise-linear function. The factor of $(g - s_0)^2$, which is very large when the stockpiled amount is small, indicates that it is important for us to buy quickly when the stockpile is low and is less urgent when it is nearly filled.

Both disruption functions, c_w for war and c_s for supply disruptions, would have the above form, though the maximum cost for supply disruptions will be less than that for war. For example, the cost of a disruption has been estimated at $1.8 billion and its probability between 0.3 and 0.7 for the 1980s [U.S. Dept. of the Interior, Bureau of Mines 1981]. We will use a probability of .05 for a supply disruption occurring in a given year. Furthermore, if we estimate the cost and probability of war at $20 billion and .005 in a year, our total weighted cost for a buying rate of 5 million lbs/yr is about $33 million. This cost must be balanced against the costs of buying too much and *not* having a crisis.

Cost of Accelerated Purchasing

The upper limit on the amount of cobalt that can be purchased at any time is computed from the social cost of making that purchase. We define the *social cost* as the extra amount that U.S. companies have to buy to get their cobalt because the government stockpile purchasing has created an artificial demand for cobalt. To determine this social cost, we need to know the dependence of price on demand. In economics, this usually is termed the *price elasticity of demand* and is defined as

$$-\eta = \frac{P}{D \frac{dP}{dD}},$$

where η is the elasticity, P the price, and D the demand.

Solving this differential equation for P gives us

$$-\eta \frac{dP}{dD} = \frac{P}{D}, \qquad \frac{dP}{P} = -\frac{1}{\eta} \frac{dD}{D}.$$

Integrating provides

$$\ln P = -\frac{1}{\eta}\ln D + C, \qquad P = CD^{-1/\eta}.$$

In our application, we want to know what happens to the price when demand is increased. Thus, we need an expression for P if D is replaced by kD (note that we use P' here for the new price, not for the derivative of the price):

$$P' = C(kD)^{-1/\eta} = k^{-1/\eta}CD^{-1/\eta} = k^{-1/\eta}P.$$

With this equation, the increase in price due to stockpile demand can be calculated: If the demand is increased by 20%, then the new price is $1.2^{-1/\eta}$ times the old price. Consequently, the social cost of buying the cobalt for the stockpile is the current U.S. demand times the difference in price $(P' - P)$. One point needs to be made here: The social cost of buying is predicated on the assumption that the original prices is the price without any stockpile buying. Thus, if we are in the second year of our stockpiling spree, the price already reflects earlier buying, and to calculate the social cost requires determining what the market price would have been if cobalt had not been bought for the stockpile.

For example, let us assume that we wish to purchase 5 million pounds of cobalt for the stockpile in 1984, at a price of \$5.10/lb. Since the U.S. purchased 6.5 million pounds [U.S. Dept. of the Interior, Bureau of Mines 1983] last year (1983), we first have to correct the world price. This is done by setting $k = 0.892$, since that is the proportion of world output that was not stockpiled, based on an estimated world production of 60 million pounds. The long-term price elasticity of demand for cobalt has been found to be -0.32 [Congress of the U.S., Congressional Budget Office 1982], so

$$P = (0.892)^{-1/(-0.32)} \cdot \$5.10 = \$3.57.$$

Thus, the beginning world market price that we will use is \$3.57/lb. Now, we wish to buy 5 million pounds. This represents an increase of 9.35% over the previous world demand of 53.5 million pounds, so the price will change as follows:

$$P' = (1.0935)^{-1/(-0.32)} \cdot \$3.57 = \$4.72.$$

The social cost of this buying for the stockpile is just the price difference times U.S. demand, which for 1985 is predicted to be 25 million pounds. Thus, the cost is (25 million) × (\$4.72 − \$3.57) = \$29 million.

Contrast this figure with the \$33 million social benefit found above. Clearly, we are buying at a rate for which the social benefits outweigh the costs; in fact, we could buy at a faster rate without real damage to the U.S. economy, and we might even wish to for reasons of national security.

What would happen if the social benefit were less than the social cost of a buying rate? This would happen only if the rate at which we are buying is really too high, and that would be a clear signal that we should immediately move to a lower rate of purchase. Notice, however, that our model always begins with a linear purchase rate, and that it is possible that the costs even of this base rate are higher than the benefits. If so, then the model has indicated that the original goal is unreasonable—it hurts the economy more than it helps, even if a war is imminent!

Now, this seems to be a problem. What if war is imminent? Should we buy at a rate that the model shows to be economically damaging so that we can survive the war? Or is the model wrong and the rate actually reasonable, given the risk of war? The answer is that if war is really that imminent, and we still find a cost/benefit imbalance, then we probably did not use the correct probability for war in the benefit analysis. We ought to use a probability very close to 1; and if we redo the analysis with such a high probability, we may find that rate is not unreasonable after all. If not, if we have put all the best data into the model and we still get a cost/benefit balance, then it is very probable that we really do have an unreasonable goal.

In general, though, the algorithm for determining how much cobalt to buy is one that maximizes expected benefit by stockpiling before a crisis while minimizing the social cost to the U.S. economy. The point of intersection of the benefits and costs is the rate at which we wish to buy cobalt, to do the most good and the least harm to the U.S. economy.

Price Fluctuations and Stockpile Management

Until now, our mathematical model has been concerned largely with the building up of the stockpile. As such, we have not explicitly dealt with the problem of price fluctuations in the market. Cobalt prices have long-range trends and short-term fluctuations. Presumably, it is desirable to purchase cobalt when the price has swung low, or before long-range pricing trends increase the cost.

The desirability of purchasing extra cobalt when the price is low, however, is already reflected in our model. When the price is low, the net social benefit of buying extra cobalt due to disruption probabilities does not change. At the same time, the net social cost of purchasing cobalt is reduced; and so extra cobalt would be bought when the price has fluctuated low simply because it would be less harmful to the economy to do so when the price is lower.

Once the stockpile has reached its goal, we are faced with a different question. We can let the stockpile sit until a crisis occurs, or we can buy and sell from the stockpile. It would be nice to have the stockpile pay for its own maintenance costs by our buying when the price is low and selling when the price is high. There also are economic benefits in keeping world cobalt prices relatively stable. Therefore, it would be advantageous to take price-fluctuation analysis into account in stockpile management.

When predicting future prices of cobalt, one looks at short-, medium-, and long-term tends in prices. This is relatively easy to do mathematically by using a least-squares fit on the previous data. Say, for instance, that you want to predict the price of cobalt n months in the future. For short-term price trends, one might look at the prices for the previous five months; for medium-term trends, for the past 20 months; and in the long term, perhaps for 100 months. By doing least-squares fit to the short-, medium-, and long-term data, one can get different predictions on the future price, along with different error bars.

If one is looking for a pricing tend over the last m months, a least-squares linear function can be found for price as a function of time (months) in the form

$$P_i = A + Bi,$$

where i is the month number, P_i is the price for that month, and the parameters A and B are given by the normal equations for a least-squares fit. We let $i = 0$ be the present month, $i = -1$ be last month, and so forth. This allows for a prediction of price n months in the future by

$$P_n = A + Bn,$$

with the standard deviation of error given by

$$\sigma_{P_n} = \sqrt{\sigma_A^2 + (n\sigma_b)^2}.$$

The usefulness of this estimate of error in the predicted price will become apparent later.

In addition, it may be useful to predict prices based on both short- and long-term trends in prices. Different predicted prices P_j, with different error bars σ_{P_j}, can be averaged together with weighting factors w_j to get

$$P_{\text{best}} = \frac{\sigma w_j P_j}{\sigma w_j},$$

with $w_j = 1/\sigma_j^2$, and

$$\sigma_{P_{\text{best}}} = (\sigma w_j)^{-1/2}.$$

Also, one can artificially weight one price prediction over another by, for example, multiplying a weighting factor w_j by 2 if one price prediction is expected to be twice as important as another.

As an example, let us take the price per pound of cobalt over the past 24 months, as shown in **Table 1**. We will predict what the price per pound of cobalt for the month of February 1985 ($i = 0$) should be, using a short-term linear fit (the last five months) and a long-term fit (the last 24 months), weighting the short-term prediction (P_s) twice as important as the long-term prediction (P_l). Using the least-squares fit equations, we get $P_s = \$5.29 \pm 0.05$ and $P_l = \$4.20 \pm 0.17$. Then, by taking the average of the two, weighted by the inverse squares of their error bars and doubly weighting the short-term prediction, we get a predicted price $P_{best} = \$5.23 \pm 0.05$. Thus, if the actual price per pound of cobalt in February is $5.10 (two standard deviations below the predicted price), it is reasonable to assume that cobalt is an exceptionally good buy this month.

Table 1.

Example prices for the last 24 months.

Month	i	$\$P_i$	Month	i	$\$P_i$
Mar 1983	−23	8.00	Mar 1984	−11	5.70
Apr	−22	7.70	Apr	−10	5.50
May	−21	7.50	May	−9	5.40
Jun	−20	7.60	Jun	−8	5.10
Jul	−19	6.90	Jul	−7	4.80
Aug	−18	7.40	Aug	−6	4.80
Sep	−17	7.00	Sep	−5	4.70
Oct	−16	6.30	Oct	−4	4.90
Nov	−15	6.20	Nov	−3	5.00
Dec	−14	5.80	Dec	−2	5.00
Jan	−13	5.40	Jan	−1	5.20
Feb	−12	5.80	Feb	0	5.10

This model allows for decisions and predictions to be made on any time scale, not just monthly. A monthly time scale was used only for purposes of illustration.

The ability to predict prices over the long, medium, and short terms allows us to take advantage of price fluctuations in the world market. Based on the expected price and the actual price, we can determine whether to buy, sell, or sit tight on our stockpile. If the situation is such that we would like to buy or sell, then we need to look at how doing so will affect the world market. This analysis is very similar to that done earlier to determine the social cost of buying for the stockpile. In this case, however, we are not interested in the cost to the U.S. economy but rather in how much which we can affect the world market price for cobalt. The central idea here is that we want to change the current price of cobalt from its extreme level to within the range of prices that our predictions consider reasonable.

If, for example, the predicted price of cobalt is $5.24 ± .05 per pound, while the actual price is $5.10/lb (more than two standard deviations away), we would like to buy enough cobalt to bring the spot-market price back within the error bounds. Thus, we know the exact price difference that we wish to achieve; what we need to determine is the amount of additional demand that we need to create to move the world price of cobalt as we desire. Previously, we showed that

$$P = CD^{-1/\eta}.$$

In this case, we want to know how demand D varies with price P. Specifically, we need an expression for D if P is replaced by kP:

$$
\begin{aligned}
D &= CP^{-\eta}, \\
D' &= C(kP)^{-\eta} = CP^{-\eta}k^{-\eta} = Dk^{-\eta}.
\end{aligned}
$$

The amount of demand that we want to create is thus

$$D' - D = Dk^{-\eta} - D = D(k^{-\eta} - 1).$$

This created demand actually can be either positive or negative. If positive, it represents an amount that should be bought, overfilling the stockpile, simply because it is a very good price for cobalt. If negative, it does not represent an artificial demand by the stockpile but rather an artificial supply, for the price is so high that money can be made by selling off a portion of the stockpiled material. One expects that this will bring the price of cobalt to more reasonable levels, and that in the future the material will be bought back at a lower price.

In our example, the predicted price for the month of February 1985 is $5.24 \pm .07$, while the actual price is $5.10. We wish to purchase enough cobalt on the spot market to bring the price up to $5.19 (within our error bars), representing a price increase of 1.76%. Putting this into our equation gives

$$D' - D = 55,000,000 \cdot (1.0176^{0.32} - 1) = 308,000 \text{ lbs.}$$

So, we need to create an artificial demand of 308,000 lbs, which we can do by contracting for that amount on the spot market.

This is a reasonable model for buying and selling the stockpile to the spot market, since any time the price falls two or more standard deviations outside the trend, it is a relatively important fluctuation. The stabilization of prices would in general benefit U.S. industries. It also could be a means for the stockpile to pay for its own maintenance.

Two important points should be mentioned about this model:

- It is not necessary to buy or sell from the stockpile every time the model so indicates. The beginning of new long-term trends will look initially like fluctuations. This would be information available to the stockpile manager and unavailable to this model. Fluctuation analysis would have to wait for a while, until the trend was established.

- Only a certain amount of buying and selling from the stockpile should be allowed, and the stockpile should never be allowed to rise or fall more than, say, 10% from the target amount. The actual amount of variation to be tolerated could be determined by the stockpile manager. However, a stockpile of the size of the U.S. cobalt reserve (more than three times annual consumption) could easily handle a certain amount of speculation in hopes of earning money and/or stabilizing prices.

Drawing Down from a Full Stockpile

The same three cases that we considered in finding the rate to build the stockpile (war, supply, and no disruption) apply to drawing it down. In each case, the manager of the stockpile must consider at what rate to decrease holdings, any lower limit to the the stockpile, and to whom to sell so as to smooth over the disruption without undue risk of running out in a future disruption. The section on price fluctuations has already dealt with the last case, managing the stockpile under normal conditions.

Mitigating the first type of disruption, that due to war, is the express intent of the Strategic and Critical Materials Stockpiling Act. In such a case, the selling rate will have been determined in part by FEMA in their estimates of how much is required for a three-year-war supply. The manager's job is to husband the supply so that defense industries are not hampered, and the supply lasts for the next three years. It is to be expected that the nonessential uses of cobalt will be decreased and that the "delayed stockpiles"—recycling, substitution, and domestic production—will begin coming online near the end of the three-year period. These factors will stretch the coverage of the defensive stockpile, and make it unnecessary for the manager to set lower limits to the amount that can be sold during the duration of the war.

A stockpile cannot be expected to eradicate all of the effects of a war on cobalt prices and availability; even if it could, it should not. There must be some incentive for developing the potential of the long-range solutions. The proper goal is to provide essential defense industries with access to this strategic metal.

The second type of disruption, that due to unstable supplying countries, is classified as an economic disruption—a problem that the defensive stockpile may not attempt to alleviate, according to current legislation. The Dept. of the Interior recognizes it as a problem that could have great cost and does have a higher probability than war [1981], and we believe that we should consider it.

One of the immediate results of the crisis in Zaïre in 1979 was panic buying by industries in an attempt to ensure that they had enough cobalt. Producer prices increased dramatically—to $25 from $5.50 in 1977—and the spot market commanded prices as high as 40%. The mere existence of a usable stockpile of cobalt would help prevent such panic buying.

Let us establish as our worst case a scenario in which Zaïre, the country that produced 47% of the world's cobalt in 1983, cuts off all production. Consider what reserves are available to make up this shortfall: Much of Africa's mine production is refined in Belgium and elsewhere, thus the metal in transit and at refineries will provide an initial buffer. Combining this with private inventories, there is a six- to eight-month supply without a stockpile. In addition, other suppliers would be expected to increase their outputs to meet part of the demand. Altogether, the stockpile would be called on to meet less than one-half of the normal domestic needs.

Notice especially that this is *normal* needs—the stockpile was intended to supply a wartime economy, with an abnormally high demand for cobalt, of three years. If the stockpile's reserves were called upon to supply half of U.S. needs for two years at the 1983 rate of use, it would use only 17 million pounds, i.e., 20% of the total stockpile goal.

The stockpile manager must decide in this case how much cobalt to keep in the stockpile in case of future disruptions and thus what limits to place on selling. This is a special case of the *ruin problem* [Feller 1966]: If orders of random size are placed at random intervals, how much of any given order can an inventory manager fill before getting an order that cannot be filled? In this special case, the cost of a war with a low inventory would far outweigh the cost of a second supply disruption. Furthermore, a disruption would increase the probability of a war and could decrease the probability of a second supply disruption. Thus, for this model we will consider only the possibility of a war following a supply disruption.

From our discussion on the social cost of war, the closer the stockpile level is to the goal, the less urgent it is to reach the goal—a 20% shortfall will cost less than 20% times the cost of war with an empty stockpile. Thus, 20% of the goal appears to be a reasonable limit. On the one hand, it is a high estimate for the need during a supply disruption; the Office of Mineral Policy and Research analysis used a two-year disruption as their worst case, and the Zaïre crisis was over in one year [1981]. On the other hand, it leaves a two-and-a-half-year supply at wartime consumption rates—which, with the two-year disruption, would be enough time for the "delayed stockpiles" to come online and extend the coverage of the defensive stockpile.

The manager could give preference to defense industries in allocating holdings during a supply disruption. However, this would probably be a mistake. It would provide a disincentive for developing long-term solutions, so that a war would find those industries unprepared.

Strengths, Weaknesses, and Testing of the Model

Perhaps the greatest strength of the model is how well it models reality. During stockpile buildup, it maximizes the benefits of early buildup while minimizing the social cost of the higher prices caused by increased demand. In the early stages of buildup, while the stockpile is still low, it calls for increased purchasing beyond the constant linear rate. Later, as the stockpile nears its target quantity, the base rate will be lower than the original rate and purchasing will slow down. This decreasing rate of purchase as the stockpile nears target is reasonable to expect from a stockpile manager.

The price-fluctuation technique is a rational way to maintain a stockpile. It helps the stockpile pay for itself and evens out price fluctuations without significantly affecting its major purpose—maintaining a supply sufficient for a three-year conventional war. This model is realistic about the limits of what a stockpile can do. It does not attempt to solve such long-range problems as pricing trends, domestic production and recycling, and long-term supply shortages. These are problems better dealt with in other ways; a strategic stockpile is at best a short-term solution to these problems.

It should be easy to write a computer program to simulate this mathematical model and run the model through a variety of tests. Although the model has been tested in the examples throughout the text of this report (the numbers in the examples closely approximate the real-life situation), it can and should be refined with a program. Such a program could experiment with a variety of probability and "social cost" functions.

This model cannot handle easily the problem of long-term pricing trends. This is not a major weakness, however. With a stockpile of this size, any attempt to capitalize on long-term trends is likely to *change* those trends, making it impossible to calculate the effects of such purchases on long-range prices.

Finally, this model is quite general. With a little work on the constants in the equations and appropriate input, it can be modified to model the management of any strategic stockpile.

Chromium and manganese are two other materials in the defensive stockpile. Each is essential to producing steel and special-alloy steels, and neither is produced domestically—over 90% of U.S. consumption is from imports. Thus, they are as liable to war disruptions as cobalt is. Further, the primary supplier for both is South Africa, for which there is concern because of racial tensions. With assessments of the costs and probabilities of the two types of disruptions, and economic analysis of the price elasticity of demand, our model could be applied to these materials too.

References

Allen, R.G. 1938. *Mathematical Analysis for Economists*. New York: St. Martins.

Feller, William. 1966. *An Introduction to Probability Theory and Its Applications*. Vol. 2. New York: Wiley.

Sibley, Scott F. 1979. *Cobalt: Mineral Commodity Profiles*. Dept. of the Interior, Bureau of Mines.

Washington, DC: Government Printing Office.

Taylor, John R. 1982. *An Introduction to Error Analysis: The Study of Uncertainties in Physical Measurements*. Mill Valley, CA: University Science Books.

U.S. Congress. Congressional Budget Office. 1982. *Cobalt: Policy Options for a Strategic Mineral*. Washington, DC: Government Printing Office.

U.S. Dept. of the Interior. Office of Mineral Policy and Research Analysis. 1981. *Cobalt: Effectiveness of Alternative U.S. Policies to Reduce the Costs of a Supply Disruption*. Washington, DC: Government Printing Office.

U.S. Dept. of the Interior. Bureau of Mines. 1981. *Cobalt: Preprint from the 1983 Bureau of Mines Minerals Yearbook*. Washington, DC: Government Printing Office.

_____. 1983. *Strategic and Critical Non-fuel Minerals: Problems and Policy Alternatives*. Washington, DC: Government Printing Office.

Veinott, Arthur F., Jr. 1965. *Mathematical Studies in Management Science*. New York: Macmillan.

Acknowledgment

This article is adapted from the authors' "The Problem of Managing a Strategic Reserve" in *Mathematical Modelling* [continued as *Mathematical and Computer Modelling*] 6 (6) (1985): 549–560. Reproduced by permission.

1986: The Hydrographic Data Problem

The table below gives the depth Z of water in feet for surface points with rectangular coordinates X, Y in yards. The depth measurements were taken at low tide. Your ship has a draft of five feet. What region should you avoid within the rectangle $(75, 200) \times (-50, 150)$?

X	Y	Z
129.0	7.5	4
140.0	141.5	8
108.5	28.0	6
88.0	147.0	8
185.5	22.5	6
195.0	137.5	8
105.5	85.5	8
157.5	−6.5	9
107.5	−81.0	9
77.0	3.0	8
162.0	−66.5	9
162.0	84.0	4
117.5	−38.5	9

Comments by the Contest Director

The problem was contributed by Richard Franke (Dept. of Mathematics, Naval Postgraduate School, Monterey, CA). His paper [1982] compares 34 approaches to this problem.

Two points on the suggested outline for papers received scant attention: testing and (especially) stability. In particular, none of the papers questioned how the depth data—all depths were given in exact numbers of feet—were arrived at: rounding down? truncation? rounding up?

Although the concept of *stability* (conditioning, robustness, sensitivity, well-posed, etc.) was introduced by the great Jacques Hadamard in 1923, it seems to have difficulty establishing itself in the undergraduate mathematics curriculum.

Reference

Franke, Richard. 1982. Scattered data interpolation. *Mathematics of Computation* 38: 181–200.

Contour Interpolation of Random Data

Chris Jacobs
John Keltner
Brian Vant-Hull
Pomona College
Claremont, CA 91711

Advisor: R.H. Elderkin

Summary

This paper addresses the problem of predicting the region of safe passage for as ship, given widely spaced and random depth soundings. We recognize the problem as one of interpolation and surface generation. We present Shepard's method of Inverse Distance Weighting (IDW) as an appropriate modeling technique and describe the history and workings of this technique. We discuss the parameters of IDW and explain their influence on the outcome. We follow with a description of the computer programs that we developed to implement this technique and to apply it to the data provided. We graph the resultant safe region. We present a possible testing scheme, followed by a discussion of possible predictors of error, and an evaluation system designed to compare one interpolation technique to another. Finally, we discuss the advantages and disadvantages of IDW.

Restatement of the Problem

The problem is to find the region of surface coordinates that have depths less than five feet. The data provided do not accurately identify this region. Therefore, some sort of interpolation technique must be applied to the data to estimate values of Z at points where depth has not been measured directly. Once estimated, these values will be used to form a contour that approximates the bottom. Then the region of depths of less than 5 ft will be identified.

Assumptions and Justification

- Depth is a *regionalized* variable. This means that the value of the variable at a point is dependent on its value in the surrounding region. Also, this effect is less pronounced for distant points than it is for local points.

- The bottom is smooth compared to the scale of our interpolation.

- The measured points that we have been provided are exact.

Note that the random (i.e., irregular) spacing of the data prohibits the use of traditional techniques of interpolation, such as splines.

These assumptions indicate the possible use of Shepard's technique of Inverse Distance Weighting (IDW).

Model Design

IDW is perhaps the most widely used computer interpolation technique in the geosciences. It first appeared in the geological journals in 1946 in the context of contour mapping and mineral-deposit extrapolation. An elaborated form of IDW formed the basis for IBM's first contouring and gridding programs of 1965. Elaborated forms of the technique are common topics in mathematical geology journals today.

The basic method of IDW consists of estimating the effects of data points on extrapolated points. The value of each extrapolated point is a linear combination of the values of surrounding data points, or, as we will refer to them, "control points." As intuition suggests, the influence of each control point is inversely proportional to the separation distance D raised to an appropriate power p (the effect of p will be discussed later). The coefficients of the linear combination must be normalized in order to avoid discontinuities in the interpolated floor. Hence, the coefficient for each control point is the product of D^{-p} and a normalization constant. Thus, the value of each estimated point is [Watson 1985]

$$Z_e = \sum_{i=1}^{n} Z_i \lambda_i,$$

where $\lambda_i = K/D_j^p$, and

$$\frac{1}{K} = \sum_{i=1}^{n} \frac{1}{D_j^p}.$$

It is the process of normalization that prevents the weighting factor from approaching infinity as values close to control points are estimated.

Since there may be thousands of control points, the modeler may opt to disregard points outside of a chosen radius. This procedure reduces the number of less important computations and also eliminates the cumulative effect of distinct control points, which could mask local trends. This procedure was unnecessary in our case, however, because of the scant number of control points.

Having presented the technique of IDWS, we now discuss its general properties. There can be no maxima or minima in the interpolated surface except near control points. Normalization of the coefficients prevents the interpolated surface from exceeding the extreme values of the control points.

The weighting of the coefficients can be varied from a local bias to a nearly equal bias by increasing or decreasing p. Local biasing increases the influence of nearby points relative to distant points, while equal biasing results in uniform influence of all points regardless of distance. Nearly equal biasing is achieved only at the price of discontinuities at the control points (see **Figure 1**). For very high values of p, the extrapolated points near each control point take on the value of the control point. The resultant surface resembles a series of polygonal plateaus around each point, connected by quickly changing boundaries. This surface is highly artificial and should be avoided.

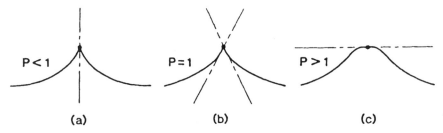

Figure 1. Behavior of the interpolated surface at the data points for three categories of the power parameter.

IDW weighting is related only to distance—it is not "ridge-preserving." More-sophisticated versions of IDW have weighting schemes biased toward control points of similar height or with similar tangent planes. These provisions ensure that an isocline between two similarly valued controls will be preserved, instead of being distorted by a nearby valley or peak.

To implement the IDW technique, we used two computer programs. The first, MAINCONT, was written just of this model. It is passed the variable p and another variable, m, indicating the coarseness of the grid on which points are to be interpolated. Using these parameters and the IDW technique, MAINCONT calculates the estimated value of the function at each node of the grid. These values are used by a second program that plots the contours; this program was written earlier by Dr. Paul Yale of Pomona College. It works by identifying nearest neighbors with values similar to the depth of the variable to be plotted and drawing lines between them. It is able to identify branch points in the isoclines and draw them correctly.

It is clear that as the grid of estimated pints is made finer, the surface will pass closer to the control points. By experimenting with different grid sizes, we found that estimating at 6-ft intervals was sufficiently fine to include all of the control points. Using similar experimentation, we found that $p = 3$ was large enough to result in appropriately pessimistic contours but not so large as to result in the unnatural plateaus discussed above.

The dangerous regions to the ship are bounded by the 5-ft isocline. The critical isoclines predicted by this model are shown in **Figure 2a**. A three-dimensional view of the surface is shown in **Figure 2b**.

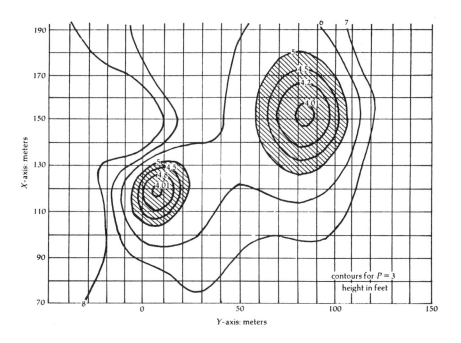

Figure 2a. The critical isoclines as predicted by the model.

Verification and Testing

The random distribution of our control points precludes any simple theoretical discussion of error; for IDW, only empirical error models were found in the literature. To test the accuracy of our results, we propose to revisit the region that we have attempted to model, to collect supplemental data. They should be collected at regularly spaced intervals, independent of the

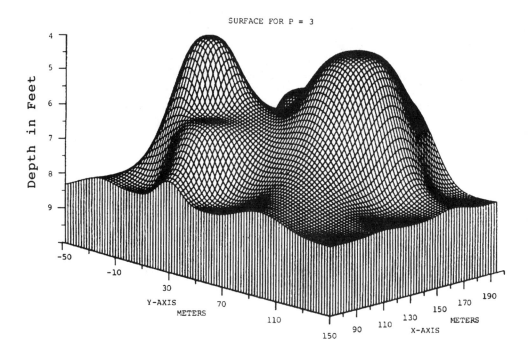

Figure 2b. An alternative method of graphing the surface.

original control points, and the root-mean-square (RMS) error should be calculated. The RMS error, as a percentage of the total range of data, can be used as a measure of the accuracy of our model. However, we could not locate standardized error levels for surface interpolation.

With these results in hand, graphical relations can be found between error and control-point density, slope, weighting exponent, and type of terrain. The intent is to produce results that can be applied to future modeling with similar techniques.

Probable error, as opposed to the actual error, can be related to local slope. This has been understood since the work of Koppe in 1905 [Yoeli 1984]. His much-used empirically derived result is

$$m_k = A \tan \alpha + B,$$

where m_k is the RMS error in depth, $\tan \alpha$ is the terrain slope at the measurement, and A and B are constants that must bed determined experimentally for each project (combination of modeling technique and terrain). However, similar projects often have the same A and B.

We expect error in depth to be somehow inversely proportional to control-point density. But the varying slopes of terrain often make a direct measure of error as a function of control-point density untenable. The Koppe error-slope formula solves this problem by permitting us to scale error measurements to the slope of each subarea within the modeled region.

The most useful error function for the modeler is the dependence of RMS error on the exponent p; the optimal value for p can be found by minimizing the RMS error. Since different types of terrain have differently shaped contours, characteristic exponents might be found for specific terrains. For example, the smooth contours of sandstone suggest a "smooth" exponent in the region of 2 or 3, while the square shape of granite suggests a higher exponent.

How is error in depth related to error in contour-line placement? From **Figure 3**, it follows that for an error m_k in depth, the contour error m_p at that point is

$$m_p = m_k \cot \alpha.$$

This equation allows us to estimate the confidence of isoclines derived from the interpolated surface.

Our discussion suggests the possibility of using a standardized test to evaluate the interpolative validity of various techniques. This test could include a standardized mathematical surface, with small-wavelength features as well as long-range trends, to which the models can be applied.

Discussion

The failures of this model are fairly limiting. It will not allow the estimated surface to extend beyond the range of the control points. It is not "ridge-preserving" in the sense that highly similar control points are weighted equally with dissimilar control points. Also, this model can supply neither an analytical estimate of its error nor confidence intervals on its predictions.

On the other hand, the major virtue of this model is its intuitive attractiveness. It is very simple to understand and implement. It is also fairly easy on computing time and does not require sophisticated software packages. These characteristics make IDW a highly usable technique that can be applied in a wide range of situations—as evidenced by its popularity in the geosciences.

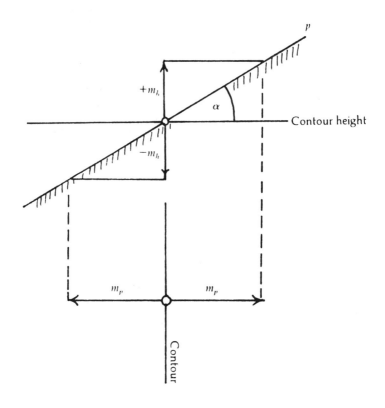

$$m_p = \frac{m_k}{\tan \alpha} = \frac{\pm(A + B \tan \alpha)}{\tan \alpha} = \pm(A \cot \alpha + B)$$

Figure 3. Relation of m_p to m_k (from Yoeli [1984, 289]).

References

Braile, L.W. 1978. Comparison of four random to grid methods. *Computers and Geosciences* 4: 341–349.

Gordon, W.J., and J.A. Wixom. 1978. Shepard's method of "metric interpolation" to bivariate and multivariate interpolation. *Mathematics of Computation* 32: 253–264.

Harbaugh, J.W., and D.F. Merriam. 1968. *Computer Applications in Stratigraphic Analysis.* New York: Wiley.

Lancaster, P., and K. Salkauskas. 1981. Surfaces generated by moving least-squares method. *Mathematics of Computation* 37: 141–158.

Matheron, G. 1963. Principles of geostatistics. *Economic Geology* 58: 1246–1266.

Olea, R.A. 1974. Optimal contour mapping using universal kriging. *Journal of Geophysical Research* 79: 695–702.

Royle, A.G., F.L. Clausen, and P. Frederickson. 1981. Practical universal kriging and automatic contours. *Geoprocessing* 1: 377-394.

Spath, H. 1974. *Spline Algorithms for Curves and Surfaces.* Winnipeg, Manitoba: Utilitas Mathematica Pub., Inc.

Walters, R.F. 1969. Contouring by machine: A user's guide. *American Association of Petroleum Geologists Bulletin* 53: 2324–2340.

Watson, D.F., and G. M. Phillip. 1984. A refinement of inverse distance weighting interpolation. *Geoprocessing* 2: 315–328.

Yoeli, P. 1984. Error bands of topological contours with computer and plotter. *Geoprocessing* 2: 287–298.

Acknowledgment

This article is adapted from the authors' article of the same title in *Mathematical Modelling* [continued as *Mathematical and Computer Modelling*] 7 (4) (1986): 577–583. Reproduced by permission.

Practitioner's Commentary: The Outstanding Hydrographic Data Papers

Ernest Manfred
Dept. of Mathematics
U.S. Coast Guard Academy
New London, CT 06320

As was stated in one of the four Outstanding papers, this is really a problem in numerical analysis rather than in navigation. The paper cited by Prof. Fusaro in his report on the competition, by the problem contributor Richard Franke, is an excellent reference for the kinds of numerical analysis that can be brought to bear on this problem.

The danger of a purely numerical analysis approach is that one may develop a model that is unrealistic. Even a simplistic approach to this problem, with an hour or two of hand calculation, will yield virtually the same region to avoid as described in the Outstanding papers; but there is no way to be sure about the region. One should not have greater "faith" in another model just because it uses more sophisticated mathematics, offers its results in fancy computer-graphic form, or involves a lot of effort.

Models are nice, but one should not put too much emphasis on them. I am reminded of a warning that I saw recently on a piece of software: "This software should not be relied upon in connection with any application in which injury to people or damage to property may result."

What do those who navigate do in practice? They follow nautical charts. And what do the chartmakers do? The U.S. organization responsible for chart development is the National Oceanographic Service. Staff there do not use models to make charts; if they are not certain of the nature of a region, they will take as many data points as necessary to get the correct topography.

Another organization that produces oceanic contour maps is the Sea Beam Facility, managed and operated by the Northeast Consortium for Oceanographic Research (which includes the University of Rhode Island, the Lamont-Doherty Geological Observatory, and Woods Hole Oceanographic Institute). Researchers there are concerned with plate tectonics in the ocean (often at great depths), not with navigation. In some instances, they will use a model to create a contour map, and they do use the spline-fitting described in one of the Outstanding papers.

1987: The Salt Storage Problem

For approximately 15 years, a Midwestern state has stored salt used on roads in the winter in circular domes. **Figure 1** shows how salt has been stored in the past. The salt is brought into and removed from the domes by driving front-end loaders up ramps of salt leading into the domes. The salt is piled 25 to 30 ft high, using the buckets on the front-end loaders.

Recently, a panel determined that this practice is unsafe. If the front-end loader gets too close to the edge of the salt pile, the salt might shift, and the loader could be thrown against the retaining walls that reinforce the dome. The panel recommended that if the salt is to be piled with the use of loaders, then the piles should be restricted to a maximum height of 15 ft.

Construct a mathematical model for this situation and find a recommended maximum height for salt in the domes.

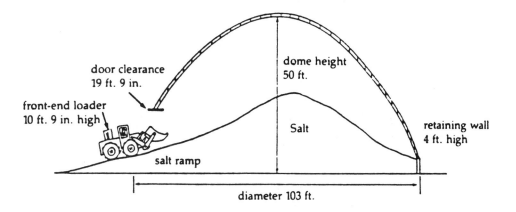

Figure 1. Diagram of a salt storage dome.

Comments by the Contest Director

The problem was contributed by Maynard Thompson (Mathematics Dept., University of Indiana, Bloomington, IN), and Indiana is the state referred to. Only 31 of 156 papers tackled this problem, which is analytically more difficult than the other 1987 problem. The two Outstanding teams, from the University of Colorado and Moorhead State University (MN), allowed a safe piling to (at least) 21 ft—a considerable increase over the panel's recommendation of 15 ft—resulting in 40% more capacity for salt.

The Salt Problem—Making a Mountain Out of Molehills

Daniel Quinlan
Jon Stoffel
Chris Sweeney
University of Colorado
Denver, CO 90202

Advisor: W.L. Briggs

Summary

Given the safety-height recommendation of the panel, we find the maximum height for the salt pile should be approximately 21.7 ft.

We developed a model employing a loading strategy to obtain a function of three parameters for maximum height. We determined that the function is stable around the solution. Variation of 10% among the three parameters produced only an 8% variation in the solution.

We implemented a computer model to generate three-dimensional images of the salt pile, to calculate volumes, and to reflect our loading strategy.

Summary

A Midwestern state must store salt for the winter. Historically, the salt has been stored in circular domed buildings using front-end loaders. The salt has been amassed to heights of 25–30 ft. Recently, a panel determined that it was unsafe to pile salt above 15 ft using front-end loaders.

We interpret the problem as finding a method to maximize the amount of salt stored (consequently its piled height) while maintaining the "absolute" safety recommendation of the panel.

Assumptions

- The roof of the dome will not support any side load.

- The retaining walls will support any 4-ft side load.

- The height of the pile of salt is measured from the floor of the dome.

- The panel's recommendation is an "absolute safe" limit that allows the front-end loader to traverse any point on the pile's surface without sliding.

Definitions

The *critical* (or *slip*) *angle* of a noncohesive granular substance (salt) is the maximum angle, as measured from the horizontal, that the substance will slide when piled. For example, when a mass is placed near a slope with critical angle, it causes the material to flow, taking the mass with it, until a new critical angle is achieved. This is the *critical angle under load* and is dependent on the mass. Therefore, there are two notable angles.

For a front-end loader, there is a *radius of reach* for the bucket.

Analysis

One type of analysis, the micro view, is to model the physical parameters of the salt, the mass of the loader, and how the salt behaves with vibration and varying forces of shear and stress. Alternatively, the macro view, while likely to be less rigorous, will be shown to depend on only three parameters:

- the slip angle (in degrees) of salt, which we denote by α;

- the slip angle (in degrees) of the front-end loader on salt, which we denote by β;

- the radius of reach of the front-end loader, which we denote by r.

We feel that these three parameters can either either derived or—better yet—determined experimentally. Further, this simplification allows us to concentrate on the method for piling salt to achieve the maximum height while retaining the panel's recommendation for "absolute" safety.

Design of the Model

First, we analyze the front-end loader. We impose a coordinate that we have called the radius of reach, which defines the distance that the loader's bucket can reach while dumping. This coordinate is measured from the base of the wheels.

The loader can then make piles of salt determined by the critical angle of the salt and the loader's radius of reach (see **Figure 1**). The height of the salt piles made is then

$$h_{\mathrm{L}} = r \sin \alpha.$$

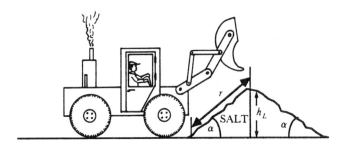

Figure 1. The front-end loader on a horizontal surface.

When the loader is on a positive slope (see **Figure 2**), the height h_{AU} of a salt pile is determined by

$$\frac{\sin(90° + \beta)}{r} = \frac{\sin(\alpha - \beta)}{h_{AU}},$$

so that

$$h_{AU} = \frac{r \sin(\alpha - \beta)}{\cos \beta}.$$

Figure 2. The front-end loader on a positive slope.

Conversely, if the loader is on a negative slope (see **Figure 3**), the height h_{AD} of a salt pile is determined by

$$\frac{\sin(90° - \beta)}{r} = \frac{\sin(\alpha + \beta)}{h_{AD}},$$

so that

$$h_{AD} = \frac{r\sin(\alpha + \beta)}{\cos \beta}.$$

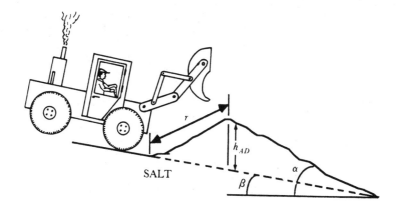

Figure 3. The front-end loader on a negative slope.

Now that we understand the limits of the loader, we next analyze the salt pile within the storage dome. The "absolute" safe condition is shown in **Figure 4**, where β is the critical angle of the loader on salt. By our assumption, the loader may go anywhere upon the conical surface. The angle β is given by

$$\beta = \arctan\left(\frac{15 - 4}{51.5}\right) = 12.06°.$$

Further, since heights of 25–30 ft are possible, and assuming a value for h_{AU} of approximately 5 ft (see **Figure 5**), we have

$$\alpha = \arctan\left[\frac{25 - 4}{100 - \frac{25-4-5}{\tan 12.06°}}\right] = 39.9°.$$

This value leads to a radius of reach for the loader of

$$r = \frac{h_{AU}\cos \beta}{\sin(\alpha - \beta)} = 10.5 \text{ ft.}$$

Since r is 10.5 ft, it is reasonable to expect that h_{AU} is approximately 5 ft, or $r/2$. These values are the ones that we use to determine the maximum height of salt.

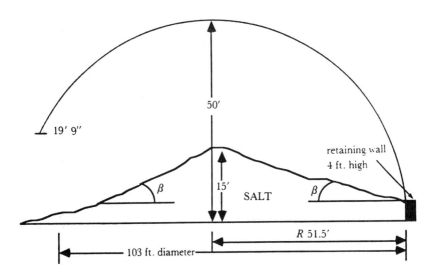

Figure 4. The salt pile.

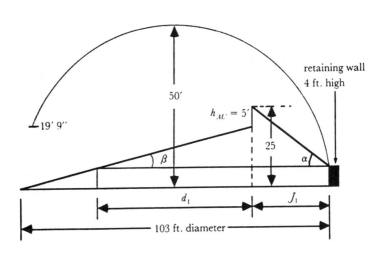

Figure 5. Determining the values of the parameters.

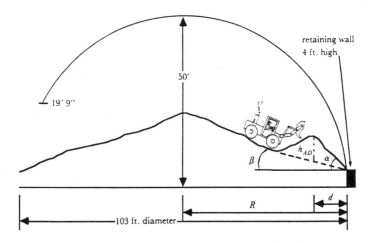

Figure 6. The panel's recommendation for the salt pile.

We believe that **Figure 6** represents the panel's recommendation and is a safe surface.

Additional storage is possible if the the loading operation follows the following procedure:

- Fill the dome until the "absolute" safe condition is met.

- Add salt on top of the safe surface, starting opposite the door and working around the periphery of the dome. The driver is instructed never to drive on the added piles.

- New deposits are positioned against the previous loads.

- To ensure stability, the loader should drive along the gradient. To avoid turns on the incline, all turns may be made at the top of the safe surface.

- Finally, the loader backfills the route up the incline to the door.

If this simple procedure is followed, more salt may be stored in the dome while maintaining "absolute" safety. The maximum height (in ft) of the salt in a dome of radius R is then

$$
\begin{aligned}
h_{\max} &= 4 + \frac{\tan\beta[2R - r\sin(90° - \alpha)\cos\beta]}{r\sin(90° - \beta)} + \frac{r\sin(\alpha + \beta)}{\sin(90° - \beta)} \\
&= 4 + 2R\tan\beta + r\sin\alpha.
\end{aligned}
$$

Using the values obtained before, we get

$$
h_{\max} = 21.7 \text{ ft}.
$$

The final cross-section of the salt mound is illustrated in **Figure 7**.

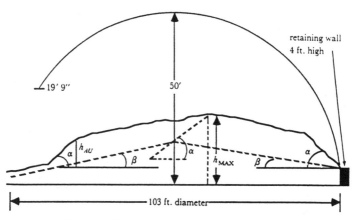

Figure 7. Final cross-section of the mound.

Testing and Evaluation of the Model

Testing of a mathematical model implies some method to check its validity. Since safety is a prime concern, we recommend that a realistic test be conducted. This would involve building a pile of salt in an open area, where the front-end loader could slide freely. With the values that are experimentally obtained, we can use the equations above to generate a three-dimensional surface (see **Figure 8**). Moreover, this surface can be used for demonstration and instruction of the salt-loading technique that we outlined.

To approximate capacity of the salt pile, we applied Simpson's rule to integrate the surface numerically. For the values in our example, the volume is 3,892 cu yd.

Strengths and Weaknesses

The strength of this model is its simplicity. If we had to explain to a layperson how we arrived at our recommendation, we could illustrate it without difficulty.

There are three specific parameters required for the use of the model, and these may be identified easily.

The weakness of this model is also its simplicity. In using only macro parameters, we may have overlooked a crucial point.

Our sensitivity analysis yields

$$\Delta h_{\max} = (\sin \alpha)\Delta r + (r \cos \alpha)\Delta \alpha + (2R \sec^2 \beta)\Delta \beta.$$

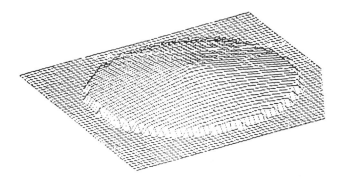

Figure 8a. Safe surface.

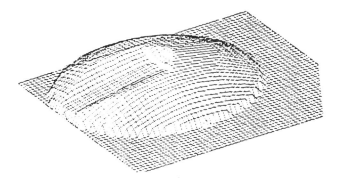

Figure 8b. Second layer.

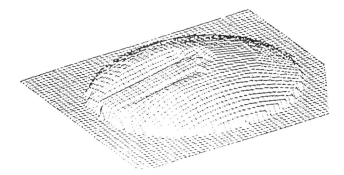

Figure 8c. Back fill.

Using the derived values for α, β, and r, and letting each of them vary by 10%, then

$$\Delta h_{\max} = \pm 1.8 \text{ ft},$$

or approximately 8%. Given these values, it appears that our method is stable.

References

Scott, Ronald F., and Jack J. Schoustra. 1968. *Soil: Mechanics and Engineering*. New York: McGraw-Hill.

Sokolovskii, V.V. 1965. *Statics of Granular Media*. London: Pergamon Press.

Taylor, A.B. 1986. *Mathematical Models in Applied Mechanics*. Oxford: Clarendon Press.

Acknowledgment

This article is adapted from the authors' article of the same title in *Mathematical Modelling* [continued as *Mathematical and Computer Modelling*] 9 (10) (1987): 747–764. Reproduced by permission.

Practitioner's Commentary: Salt Storage Done Improperly, Safety Panel Says

Kyle Niederpruem
Staff Writer
Indianapolis Star
November 29, 1986, p. 33
Reprinted with permission of the *Indianapolis Star*

Piles are Too Large, Could Shift Dangerously, State Told

A violation cited by the Indiana Occupational Safety and Health Administration will result in a change in state highway policy and may force a $600,000 purchase of conveyors for loading salt.

"We're telling our people—don't climb a mountain to put the salt in," said Donald W. Lucas, deputy director of operations for the Indiana Department of Highways.

The storage practice in Indiana's 80 salt domes has been the same since the early 1970s.

Employees drive on ramps of salt leading into the domes and pile salt up to 25 or 30 feet with buckets attached to 10- or 15-ton front-end loaders.

"Even though we haven't had an accident history, it's probably not a good idea to use front-end loaders to pile salt any higher than 15 feet," Lucas explained.

Several employees, whose names are not part of the IOSHA record, made informal complaints about the practice concerning one of the 50-foot-high domes in the Crawfordsville District.

"In piling or storing salt in this manner, if the end loader gets too close to the edge of the pile, the salt can shift," says the IOSHA report, which followed two inspections in July and August.

The shifting pile of salt could throw the loader against retaining walls that reinforce the dome.

There were at least two incidents in Crawfordsville involving front-end loaders slipping on salt piles, but neither resulted in injury to the operator nor damage to equipment, said Larry R. Vaughan, district maintenance engineer.

Mechanical conveyors, similar to ones used for fertilizer and grain, may be a partial answer. The conveyors, however, cost about $30,000 apiece.

And because salt is so corrosive, a stainless steel model may last only one season.

The department is discussing the purchase of up to 20 conveyors, for $600,000, that could be transported by truck to several districts and shared.

IOSHA's citation came as a surprise to highway officials and one New Jersey manufacturer. According to state law, an employer must provide a place to work free of hazards likely to cause injury.

"I don't think the citation is worthy of the fuss," said David M. Wilcox, district engineer in Crawfordsville.

"Everybody agrees. Some sort of mechanical filling process would be great. Why change it for the sake of spending money. Every dollar spent needlessly takes it away from road service. But I'm not in a position to argue."

Only one of the district offices for the department has been cited, even though the same loading method is practice around the state. But as one official noted, "If it's a problem for Crawfordsville, it's a problem for everybody."

Lucas said conveyor were purchased for another district about two years ago when highway officials were attempting to address the safety factor without any mandates from IOSHA.

By citing budget constraints, the highway department was able to have the violation abated until May 26, 1987. The domes have already been loaded with 240,000 tons of salt for this winter season.

Dome Corporation of America, a New Jersey company, is just one of the suppliers for Indiana. Dome Corp. has sold 2,000 salt domes nationwide and only 1% have conveyor systems for loading.

"Some of them have gone with conveyors. It's recent. And it's an expensive proposition. You're dealing with a corrosive material," said Stan Milic, president of Dome Corp.

"People have not been hurt. Common sense will tell you—you can only climb so high," he added. Indiana is the only state Milic is aware of that has been cited for a safety problem.

Robert L. McCreary, commissioner of the Indiana Labor Department, said the highway department has been "agreeable" to making the needed changes in storage policies.

"I don't think there will be a problem. They're trying to get this resolved. It's a lot of money, but I'm convinced we'll work it out," said McCreary.

1988: The Drug Runner Problem

Two listening posts 5.43 miles apart pick up a brief radio signal. The sensing devices were oriented at 110° and 119°, respectively, when the signal was detected (see **Figure 1**); and they are accurate to within 2°. The signal came from a region of active drug exchange, and it is inferred that there is a powerboat waiting for someone to pick up drugs. It is dusk, the weather is calm, and there are no currents. A Small helicopter leaves a pad from Post 1 and is able to fly accurately along the 110° angle direction. The helicopter's speed is three times the speed of the boat. The helicopter will be heard when it gets within 500 ft of the boat. This helicopter has only one detection device, a searchlight. At 200 ft, it can just illuminate a circular region with a radius of 25 ft.

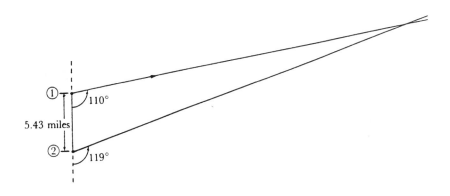

Figure 1. Geometry of the problem.

- Describe the (smallest) region where the pilot can expect to find the waiting boat.

- Develop an optimal search method for the helicopter.

Use a 95% confidence level in your calculations.

Comments by the Contest Director

The problem was contributed by J.A. Ferling (Mathematics Dept., Claremont McKenna College, Claremont, CA). It is a modified version of a classified military problem. Neither the original nor the modified problem has a known optimal solution.

Locating a Drug Runner: Miami Vice Style

Sean Downing
John Humanski
John Seal
Drake University
Des Moines, IA 50311

Advisor: A.F. Kleiner

Summary

We have been assigned to develop a procedure to determine accurately the region in which a suspected drug runner is located, and to find the "optimal" helicopter search method to capture the runner. We make initial assumptions regarding the specifics of the helicopter and the powerboat, the accuracy of the detection devices, the motions of the drug runner, and the specifics of the search. The probable region containing the speedboat is determined by applying statistical analysis to the accuracy of the detection devices. We investigate several possible search strategies, and the elimination of ineffective patterns results in the consideration of two promising paths. We test our model based on drug-runner psychology and the two search strategies. The development of a computer simulation of the interaction between the helicopter and the speedboat allows us to test variations in each scenario. We discuss the resulting data and present an optimal solution based on our model. The model is evaluated, and we suggest further improvements and refinements.

Assumptions and Hypotheses

The problem statement contains several unknowns that we will have to assume; however, these assumptions are either based on documented information or appear reasonable in all cases.

Helicopter Specifics

We assume that the helicopter is similar in design and specifications to the U.S. Coast Guard Sikorsky HH–3F, which are well able to reach a speed of over 200 mph [Boyne and Lopez 1984, 108, 249]. However, we assume that the helicopter's speed is a constant 180 mph (264 ft/s), since the pilot must be actively searching the ocean below. We also assume that the helicopter's searchlight is mounted underneath the cockpit, so that a full 360° "sweep zone" is possible. We further restrict this zone to a 90° region extending 45° on either side of the forward motion of the helicopter. This refinement is reasonable if we assume that the searchers can view only a region directly ahead of their path. Since we reduce the sweep zone by three-quarters, we will assume that the searchlight can illuminate this entire region continuously; that is, the beam can sweep fast enough to illuminate the entire 90° region. Since the weather is calm, we assume that the helicopter can fly safely within 20 ft of the ocean. Finally, we assume that the helicopter is sophisticated enough to allow an exact flight pattern to be followed and maintained. This seems acceptable if we assume that the helicopter has a computerized navigation system.

Speedboat Specifics

Since the problem statement specifies the drug runner's powerboat has a speed of one-third the helicopter, we assume that the boat speed is 60 mph. Furthermore, we assume that the runner's acceleration is instantaneous and that the boat maintains top speed when moving. Since the typical "go-fast" motorboats often used by drug runners can obtain speeds of 80 mph [Shannon 1986], these assumptions appear reasonable.

Detector Accuracy

We interpret the error of ±2° as a normal distribution, centered on the given angles (119° and 110°), with standard deviation of 1°. Thus, the true values will fall within 2° of the measured value with probability of approximately 95%.

Psychology of the Drug Runner

We assume that the drug runner will remain at the original position indefinitely, as long as the helicopter is not heard. If the runner does hear the helicopter, we assume that the boat will attempt to escape by following a path perpendicular to the direction of motion of the helicopter and will move at maximum velocity. The runner's motion following the detection of the helicopter is a factor that we consider separately.

Specifics of the Search

Considering limitations in fuel and pilot fatigue, we assume a 6-hr time limit on the search. That is, if the runner is not sighted after 6 hrs, the search will end and the runner will have escaped. Certainly, the drug deal would have been completed within this time limitation. We also assume that the helicopter engages the search at a specified point by flying from the coast at an altitude of 600 ft, beyond the hearing range of the runner, and then descending to water level. In this manner, the helicopter may begin the search at any arbitrary point while maintaining the same starting conditions. Finally, if the powerboat is within sight of the helicopter, we assume that the drug runner will be apprehended.

Brief Problem Overview

The first problem to attack is the description of the region in which the pilot can expect to find the waiting powerboat. The given information includes only the results of the sensing devices used to triangulate the possible locations of the drug runner. Since we assume that the sensing devices operate according to a prescribed statistical method, we must use some sort of probability interpretation to describe accurately the region in which the boat will be found. In any case, however, we can take the given accuracy of the sensing devices, ±2°, to calculate a first approximation of the desired region. This result will give the *minimum possible* region in which the boat lies waiting and is instructive in further analysis.

After determining the correct region in which the speedboat lies, the next problem is to devise an optimal search strategy to capture the drug runner within the assumed 6-hr time limit. We interpret "optimal strategy" as the method that will give the highest success rate in capturing the speedboat. This interpretation disregards flight time, gas expenditure, and ease in implementation; however, we feel that the best strategy should concentrate on the actual *capture* of the drug runner, regardless of any other factor.

The most important consideration in developing a search strategy is the fact that the speedboat can hear the helicopter at 500 ft, but the pilot can see the boat at only 200 ft. Thus, we have the case that any helicopter approach pattern has the highly probable result of "scaring" the speedboat away from the pickup point, without a helicopter sighting. This consideration led us to consider only "closed-curve" flight patterns; that is, ones in which the helicopter would follow a circulating path. As a result, the helicopter has higher probability to discover the drug runner if the helicopter follows an inwardly or outwardly progressing path. We will investigate two such flight patterns: perfect circles centered on the most probable location of the speedboat, and quadrilaterals following the general dimensions of the region in which the pilot expects to find the drug runner.

Another important consideration is the movement of the speedboat once the drug runner hears the helicopter. The responses could be quite diverse; however, we concluded that three such responses are most probable:

- The runner might simply continue fleeing, never going back to the pickup point, thinking that the drug deal has been scrubbed.

- The boat might move away from the pickup point, but after a certain time stop and wait indefinitely for further instructions from its contact.

- The boat might initially move away from the point, but after a certain time stop and head back to the same point to complete the drug exchange.

Each of these three scenarios certainly has strengths and weaknesses; but without more detailed information on the landscape and other factors, we include all three for completeness.

Problem Analysis

Defining the Region

Since we are assuming 95% confidence that the sensing devices used to locate the speedboat are accurate to 2°, we may use the *interval error method* to determine the extent of the possible location of the waiting speedboat. That is, we may examine the maximum and minimum deviations in each sensor to determine the shape and size of the resulting region. The result of this basic trigonometric procedure are shown in **Figure 1** and magnified in **Figure 2**. Although the region is quite extended in two directions, we note that the *most probable* location of the drug runner is at the intersection of the 110° and 119° sensor lines from the listening posts. This is a result of our assumption that the true angle varies on a normal distribution centered at the given angle. Thus, for further computations and analysis, we choose to relocate our origin from the second listening post to the point of greatest probability density (see **Figure 3**).

Developing Optimal Search Pattern

Central to the problem is the idea of the "catch zone," the region about the helicopter in which the runner will be caught. The problem states that at an altitude of 200 ft, the helicopter's searchlight will illuminate a region 25 ft in radius. However, if the helicopter maintains an altitude of about 20 ft, the effective search zone increases to a radius of almost 200 ft (see **Figure 4**). Thus, any object within 200 ft of the helicopter and within the beam of the searchlight will be detected. Essentially, then, except for the initial "drop in" at the start of the search, we have reduced the problem to two dimensions.

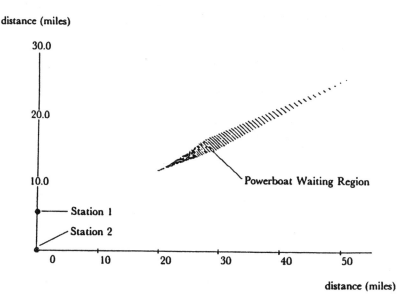

Figure 1. The powerboat waiting region as determined by the interval error method.

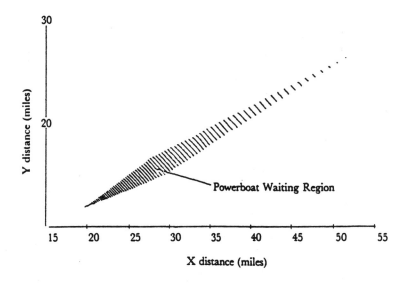

Figure 2. Magnification of **Figure 1**.

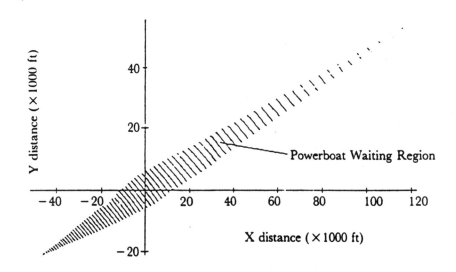

Figure 3. As in **Figure 2,** but with origin relocated to point of greatest probability density.

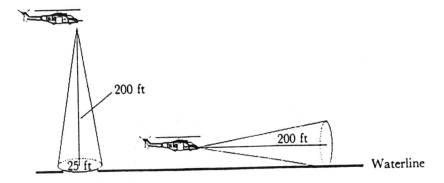

Figure 4. Different ways to use the searchlight.

The searchlight is allowed to sweep out a continuous path 90° wide, creating a "sweep zone" that represents the extent of the searchers' vision (see **Figure 5**).

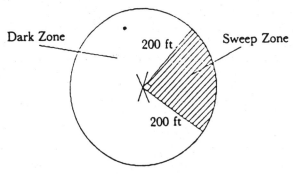

Figure 5. Extent of the searchers' vision.

The remaining portion of the 200-ft-radius circle, the "dark zone," represents that portion *not* instantly swept out by the beam. However, since the helicopter is flying forward at 264 ft/s (180 mph), this dark zone will in effect be searched, since the helicopter has already moved through the region. Thus, we have a "catch region" 283 ft across, which guarantees the drug runner's capture if the speedboat lies within the region (see **Figure 6**).

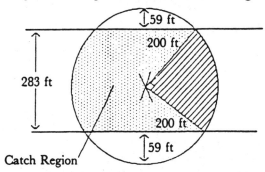

Figure 6. Geometry of catch region.

Although there are two additional 59-ft portions of the pilot's 200-ft radius outside the catch region, we cannot envision a situation in which the boat would be within these sections, since the drug runner has a 300-ft hearing advantage and would have moved out of the helicopter's range. Effectively, then, if the speedboat lies anywhere within a 200-ft radius of the helicopter , then either the drug runner is in the 90° sweep zone and will be apprehended, or else has already been captured.

An important factor in developing the search pattern is the difference between the drug runner's 500-ft hearing range and the helicopter's 200-ft sight zone. It is likely that on any given pass, the drug runner will both hear the helicopter and be outside the catch region, thus initiating the boat's movement away from the helicopter's path. Our search pattern, then, must take into account the probable result that the boat will be moving in some fashion. We initially considered many search patterns but found that several would not be feasible. One possibility was some sort of spiral, such as the spiral of Archimedes ($r = a\theta$) or the exponential spiral ($r = ae^{\theta}$), which continually curve away from the center. Unfortunately, any such spiral became impractical due to difficulties in establishing parametric equations (as a function of time) that would allow constant velocity along the length of the spiral. Other possibilities included any set of straight line segments or sine waves adjusted to the boundaries of the search region. These proved impractical, however, since initial calculations showed that it is impossible to cover the *entire* region within the 6-hr time limit. Thus, the best strategy would be to follow a path that would both cover the region of highest probability (the origin) and also "loop" back upon itself, increasing the likelihood of intercepting the boat.

The two patterns that we examine are shown in **Figures 7** and **8**, superimposed on the powerboat waiting region.

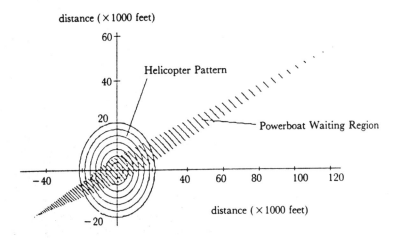

Figure 7. Circular Method of search.

In the Circular Method of **Figure 7**, the helicopter travels along circles centered at the most probable point in the region. In the Quadrilateral Method of **Figure 8**, the helicopter travels along a pattern geometrically similar to the boundaries of the powerboat waiting region. We will investigate the difference in capture rates for varying distances between successive patterns and varying starting points. We will also examine the effects of changing the direction of travel ("inward" or "outward") along the given pattern. (Note that the distances between successive patterns is quite large

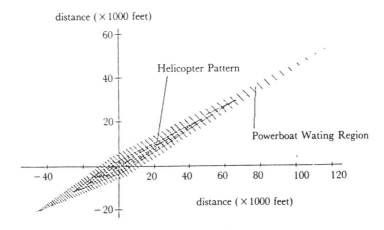

Figure 8. Quadrilateral Method of search.

in he figures; this is for illustration only, since for the practical search one would want no more than perhaps 2,000 ft between successive patterns.) Notice that in the circular pattern, the helicopter travels significantly beyond the limits of the powerboat waiting region, while the quadrilateral scheme it remains entirely within the region. This difference will likely manifest itself in the three different drug-runner reactions described above, in that the Circular Method will likely perform best for the drug runner's flee response, while the Quadrilateral Method will be best for the cases in which the runner stops and waits or returns to the pickup point.

Model Design

The heart of our model is a set of computer routines that simulate the interaction between the helicopter and the speedboat according to the assumptions stated previously. The computer routines are flexible enough to allow us to investigate a variety of different factors in the interaction and perform repeated application of these factors in a large number of test cases.

Our computer routines reduce to a $3 \times 2 \times 4 \times 2$ model, in which we examine the effects of the following four factors on the capture rate:

- *Drug runner's response to hearing the helicopter.* The runner is assumed to exhibit one of the following three responses:

  - The boat moves away from the sound of the helicopter and neither stops nor returns to the initial point.
  - The boat initially moves away from the sound but stops 5 mi away and waits for further instructions.

    – The boat moves away from the sound, stops 5 mi away, but returns to the initial pickup point after waiting 10 min, provided the runner does not hear the helicopter again.

• *The type of flight path followed by the helicopter.* We include consideration of the two types of "closed paths" mentioned above, the circular and quadrilateral schemes. In either method, the helicopter will navigate complete loops.

    – In the Circular Method, the helicopter begins and ends each circle at a specified distance along the x-axis, and then travels directly along the axis before beginning the next circle (see **Figure 7**).

    – In the Quadrilateral Method, the pilot begins and ends each loop at a specified distance along a line joining the most probable point (the origin) and the closest exterior corner point of the powerboat waiting region to the origin (see **Figure 8**).

• *The distance between successive loops in either the Quadrilateral Method or the Circular Method.* This parameter effectively indicates the "tightness" of the search pattern: For a large distance between each loop, the helicopter covers a larger total region but misses a significant portion of the intervening area. For a small distance, on the other hand, the helicopter covers a small region more thoroughly. Considering the helicopter's 200-ft sweep zone, we examine the success rate of four different distances: 200, 500, 1,000, and 2,000 ft.

• *The radial direction of the helicopter's path*: that is, whether the pilot follows successive loops outward from or inward toward the most probable point. Logically, the helicopter would start either at the most probable point and work outward or at an outer extremity and progress inward, since the pilot would want to "scare" the drug runner in the same direction.

    Therefore, our analysis requires the simulation of 48 different configurations, from which we will draw conclusions as to the most successful method under various constraints.

    The computer program is an interactive routine that uses basic algebraic methods to compute the flight path of the helicopter at one-second intervals throughout the 6-hr period. The program then determines if an interaction between the helicopter and the speedboat occurs at any point, subsequently calculating the motion of the drug runner according to the criteria established above. The process is simplistic, certainly; but it establishes an acceptable approximation to an actual search routine.

    For each of the 48 configurations, we tested the computer simulation for the same 134 starting locations of the powerboat. These 134 points were derived in the following manner. We generated random numbers according to a normal distribution centered about the two given angles, 110° and 119°, and having a standard deviation of 1°, based on the assumption that the given error of 2° represents a 95% confidence interval. Triangulation of these pairs of random angles yielded 134 starting locations within the powerboat waiting region (see **Figure 9**).

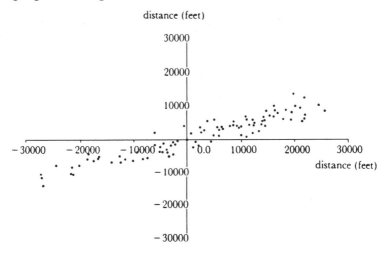

Figure 9. Search locations within the powerboat waiting region.

The sample size of 134 will result in a *p* (probability of catching the powerboat) for each of the 48 scenarios which will be accurate to within .051 at a 95% confidence level, assuming a *p* in the neighborhood of .10:

$$z_{.025}\sqrt{\frac{p(1-p)}{n}} = 1.96\sqrt{\frac{(.10)\cdot(1-.10)}{134}} = .051.$$

Problem Predictions and Results

We choose to examine the capture rate based on the three possible "psychologies" of the drug runner. That is, for each of the responses—either stopping, returning, or fleeing after the initial sound of the helicopter—we investigate the best possible search strategy. We feel that this method is most practical, since various factors such as weather and coast configuration will alter the runner's response. Using this scheme, then, our results can be applied for a variety of runner responses.

Tables 1–3 show the success rate—that is, the number of captures for the 134 initial boat locations—for the various parameters discussed above and for each drug-runner response. For example, in **Table 1**, for the Circular Method, starting at a large radius and progressing inward in steps of 500 ft, we find that the drug runner was captured in 18 of the 134 starting points, giving a success rate of .134.

Table 1.

Strategy 1: Drug runner flees pickup point and doesn't stop or return.
Number of catches (proportion of catches) for 134 random starting points.

Patterns:		Distance between concentric patterns			
		200 ft	500 ft	1,000 ft	2,000 ft
Circular	Outward	1 (.007)	14 (.104)	19 (.142)	11 (.082)
	Inward	2 (.015)	18 (.134)	22 (.164)	10 (.075)
Quadrilateral	Outward	8 (.059)	12 (.090)	7 (.052)	5 (.037)
	Inward	10 (.075)	9 (.067)	14 (.104)	14 (.104)

Table 2.

Strategy 2: Drug runner flees pickup point and stops to wait.
Number of catches (proportion of catches) for 134 random starting points.

Patterns:		Distance between concentric patterns			
		200 ft	500 ft	1,000 ft	2,000 ft
Circular	Outward	1 (.007)	13 (.097)	14 (.104)	11 (.082)
	Inward	6 (.045)	25 (.187)	23 (.172)	9 (.067)
Quadrilateral	Outward	12 (.090)	11 (.082)	7 (.052)	4 (.030)
	Inward	10 (.075)	9 (.067)	11 (.082)	12 (.090)

Table 3.

Strategy 3: Drug runner flees pickup point, stops to wait, and returns.
Number of catches (percentage of catches) for 134 random starting points.

Patterns:		Distance between concentric patterns			
		200 ft	500 ft	1,000 ft	2,000 ft
Circular	Outward	7 (.052)	18 (.134)	18 (.134)	12 (.090)
	Inward	15 (.112)	23 (.172)	21 (.157)	9 (.067)
Quadrilateral	Outward	12 (.090)	9 (.067)	7 (.052)	3 (.022)
	Inward	10 (.075)	11 (.082)	9 (.067)	14 (.104)

These tables suggest that the optimal strategy of those tested is a circular pattern with inward movement and intermediate distance (500 or 1,000 ft) between consecutive rings. This is not surprising, since a circular pattern increases the probability that the helicopter will intercept the boat should the runner have already been set in motion. An inward, as opposed to outward, pattern will tend to chase the boat toward the center of the region, rather than "out to sea," thus effectively reducing the region in which it is most likely to be caught. Patterns with small distances between each ring will cover only a small portion of the total region in the allotted time, while very large patterns leave a greater space in which the boat may be passed over. Thus, those of intermediate size should result in a higher capture ratio, as observed.

However, as the tables show, there are some exceptions to these general rules. While the Circular Method tends to be superior in most cases, at a very small distance between concentric rings, the Quadrilateral Method results in a higher catch ratio. Further, the superiority of the inward direction seen so clearly with the circular pattern is not realized when the quadrilateral search pattern is used.

The particular cell that shows the highest capture rate varies from chart to chart, so it is difficult to make a specific recommendation to the helicopter pilot. The best course of action depends on the strategy selected by the drug runner, a strategy that the pilot could not know. If we assume that the three strategies that we investigated occur with equal probability, then the probabilities of a capture are equivalent for the Circular–Inward–500 strategy and the Circular–Inward–1,000 strategy. It is also important to consider the statistical error of approximately 0.05 based on our random sample size, so choice of a clear-cut optimal strategy is further clouded.

If we consider the perspective of the drug runner, then Strategy 3, in which the boat returns to the initial spot to complete the drug deal, is by far the poorest; it is most likely to result in capture. From this point of view, there is no appreciable difference between the other two strategies.

While movement in the inward direction appears optimal in terms of our criteria—the proportion of trials that result in capture within 6 hrs—this may not be true in terms of other criteria. Because the region of highest probability of finding the drug runner is reached last in this pattern, a greater expenditure in terms of effort, helicopter fuel, etc., may be required on those trials that do result in capture.

Model Strengths and Weaknesses

The most important technical weakness in our procedure lies in our consideration of only two helicopter flight patterns. An infinite number of paths could be investigated, some quite complex. We believe, however, that our examination of the two methods (circular and quadrilateral patterns) provides a good indication of the model's success. Smoothing the circular pattern out into a true spiral would probably be more efficient, but the parametrization proved impractical. The model also is limited in its assumption about the drug runner's path following the initial sound of the helicopter. While we assume an escape path perpendicular to the tangent of the helicopter's direction, in fact the boat's motion could be quite chaotic. Similarly, our analysis considers only three drug-runner responses (flees, waits, or returns). The real-life response could be considerably more complex, perhaps even combining elements of all three. Also, we do not consider the time of flight of the helicopter from the Coast Guard station to its initial search point. This seems acceptable, however, since even the most distant starting point would be less than 31 mi from Post 1, a distance that the helicopter could cover in less than 15 min, a small portion of the total search time. Finally, our definition of "optimal" neglects the time factor involved in the search; we treat only whether the runner is caught at *any* time within the 6-hr limit. If it became necessary to minimize the search time, further analysis would be necessary.

The most important strength of our model lies in the calculation of the exact trajectories of the helicopter and the speedboat at *each second* within the 6-hr time interval. The computer routine therefore examines 21,600 points of possible helicopter-speedboat interaction, eliminating the possibility that the paths will cross undetected. Our model is successful in that it considers many possible scenarios. We examine 48 different configurations, varying flight patterns, directions, and drug runner responses. Our computer routine may be conveniently adapted to include consideration of any other helicopter flight pattern. This flexibility would facilitate future changes or improvements necessary in the scenario. Finally, the model is statistically accurate, since the initial powerboat points are chosen according to the more realistic assumption that the accuracy of the sensing devices is represented by a normal as opposed to a uniform distribution.

References

Boyne, W.J. and D.S. Lopez, eds. 1984. *Vertical Flight: The Age of the Helicopter.* Washington, DC: Smithsonian Institution Press.

Shannon, E. 1986. Running silent, running fast. *Newsweek* (27 October 1986): 94–95.

Practitioner's Commentary: The Outstanding Drug Runner Papers

Walter Stromquist
Daniel H. Wagner Associates
Station Square Two
Paoli, PA 19301

I liked the comment of the team from the North Carolina School of Science and Mathematics about not expecting the searchers to have an advanced computer when they can't even afford radar. An actual search under these conditions would almost surely use radar, so this problem is somewhat artificial. Also, the distances mentioned in the problem are somewhat unrealistic: A helicopter can be heard for miles, and a good searchlight can illuminate a boat in open ocean at a distance of a mile or more—at least well enough to make the searchers come back for a closer look.

I suspect that this problem has been contrived as a proxy for a much more realistic problem in antisubmarine warfare. For example, the target might be a lurking submarine, and the searcher might be a destroyer using active sonar. The destroyer sends out "pings," which will allow it to detect the target at a certain distance; but the pings can be heard by the submarine at a greater distance. The submarine is effectively much slower than the destroyer, because if the submarine attempts to go too fast, the destroyer's sonar operators will hear the sound of the submarine's engines. Alternatively, the searcher could be another submarine. Searches of this kind arise repeatedly, so even if the problem is artificial in the form presented, it is not irrelevant.

Problems like this are called (not surprisingly) *searcher path problems*. From a theoretical point of view, they are notoriously difficult—in fact, some versions of the problem are NP-complete. Therefore, we are not likely to find any efficient mathematical algorithm to compute a provably optimal path. In practical situations, there is no better approach than the one chosen by both of the Outstanding teams: Think of some candidate search plans, evaluate each plan on the basis of some model, and pick the best of the candidates. There is never any assurance that we have thought of the best plan.

Still, there are some analytic techniques available for dealing with these problems. I will describe some of them under these headings:

- the probability distribution;

- sweep width; and

- candidate search plans.

Then I'll comment on wake detection and other topics.

The Probability Distribution

Both teams did a fine job of describing the target's initial location. Both teams assumed (as I would have) that each of the bearings is subject to a normally distributed error with standard deviation 1°. They could therefore draw a wedge around each bearing line, extending 2° to either side of the line, such that each wedge has a 95% chance of containing the target. The intersection of the wedges is a "search area" roughly 35 mi long and 2.5 mi wide at its widest point, with area about 40 sq mi. Note that the probability that the target is actually in this area (i.e., in both wedges simultaneously) is about 90%.

Both teams recognized that the search area is really only a convenient way of representing a complicated bivariate probability distribution, and both teams found effective graphical ways of illustrating the distribution.

What if the target has been moving since the message was intercepted? In that case, the search is probably hopeless, so we may as well assume (as both teams implicitly did) that the target is stationary.

Sweep Width

We can model the searcher as a broom, sweeping a path through the search area and finding anything within the path. The shape of the broom, or *catch zone*, is unimportant; all that matters is its width, which we call the *sweep width*. It is one of the most important concepts in search theory, because it summarizes the capabilities of the sensor in a single number. If we aren't sure of the number, we can vary it parametrically and, in effect, test the sensitivity of all of our sensor-related assumptions at once. Both teams came up with numbers for the sweep width. In the simplest cases, the North Carolina team used 400 ft (the diameter of a circle), and the Drake University team used 283 ft (the width of the forward-looking 90° wedge).

Neither team really came to grips with the use of the searchlight. If the helicopter is traveling at 140 or 180 mph, then it will be impossible to swing the beam back and forth fast enough to find all targets in a 283-ft or 400-ft path. Even if it takes only 2 sec to sweep from left to right and back to the left again, the helicopter will have traveled at least 400 ft, leaving the left side of the path uncovered for most of this time. For part of the cycle, the searchlight will be centered, making the sweep width nearly zero. Perhaps the best strategy is to keep the beam pointing to one side, at right angles to the path; that would guarantee us a sweep width of 200 ft.

Both teams devoted substantial effort to modifying the sweep width based on the target's evasion strategy. This led to some elegant geometry, but it was probably a mistake, given the distances and speeds in the problem. Once the helicopter has come close enough to be heard by the boat, it will cover the 500 ft between itself and the boat in less than 3 seconds, giving the boat no time to react. It is better to assume that the boat is stationary, and that any evasion strategy takes effect only after the first encounter is complete. This leaves the sweep width unaffected by the evasion strategy, considerably simplifying the problem.

If we ignore the "scare zone," then we can use the sweep width to obtain a simple estimate of the ultimate success probability of the search. Multiply sweep width by searcher speed to get *sweep rate*, in units of area covered per unit time; then multiply by the time available to get the total area that can be covered. Using the North Carolina team's numbers, the calculation is even easier: The length of the search path is assumed to be 225 mi., so that the area that can be covered is (400 ft) × (225 mi) = 17.0 sq mi. This is less than half of the search area; but by covering the 17 sq mi of highest probability, we can obtain a success probability of 50–60%. This is almost without regard to the path chosen: All that is necessary is that we stay in the high-probability area and avoid overlapping previously searched territory.

But we cannot ignore the scare zone. A better model is that the path of the broom is 1,000 ft wide, as in **Figure 1**.

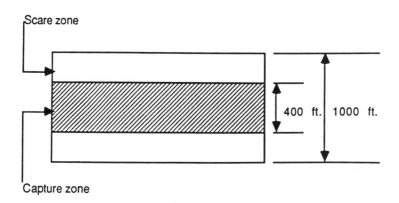

Figure 1. Model of the searchers' path.

Outside the capture zone (say 400 ft wide) is a space in which the target will hear the helicopter and presumably escape before the helicopter returns. With a sweep width of 1,000 ft, we can cover 43 sq mi in the time allowed; but even if the target is in the area covered, we have only a 400/1000 = 40%

chance of capturing it. It is better to use the North Carolina team's idea of spacing the tracks as in **Figure 2**, so that one side of the scare zone overlaps previously covered territory. Now our effective sweep width (sweep width in new territory) is 700 ft, so we can cover about 30 sq mi, and the chance of capturing the target given that it is in the covered area is 400/700 = 57%. More accuracy than this requires a simulation.

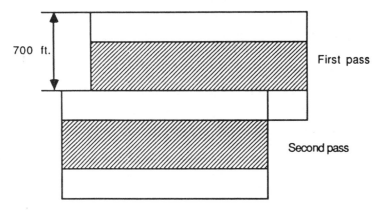

Figure 2. Overlapping paths.

Candidate Paths

The argument from sweep width shows that the exact path chosen makes little difference, as long as the path stays within the highest-probability area and the track spacing is such as to avoid unwanted overlap.

If the target is not expected to escape (or if it is expected to return immediately, which amounts to the same thing), then the appropriate track spacing is equal to the sweep width. This avoids overlap of coverage while still packing the tracks as tightly as possible into the high-probability area. The Drake University team's simulation results bear this out: In the cases with an escape-and-return strategy, the smaller track spacings were most effective. The results also show that track spacings of 500–1000 ft are most effective if the target is prone to escape permanently; this is consistent with the plan of **Figure 2**.

Aside from these considerations, any preference for one path over another has to be based on something more subtle than our simple sweep-rate model. For example, both teams were quite ingenious in designing paths that maximized the probability of catching the target in the course of its escape. This is good thinking, but there is no way short of a simulation to test the effectiveness of their ideas.

Another consideration involves timing. There is the constant risk that the target will complete its business and leave the area. Therefore, it is better, all other things being equal, to search in the highest-probability areas first. This is an argument for the "outward" patterns (in the Drake University team's scheme) that isn't addressed in their simulation.

Wake Detection and Other Topics

Some readers may be surprised to learn that the North Carolina team's assumptions about the persistence and detectability of the boat's wake are very reasonable. A fast-moving power boat in open ocean may indeed leave a wake that can be detected from the air, even by searchlight, for as long as half an hour. The wake makes itself more visible by stirring up luminescent plankton. Of course, this depends on the weather and on the size and speed of the boat.

The North Carolina team also proposed a very clever method of taking advantage of the wake: by using the same track spacing as in **Figure 2**, they ensure that if the boat is "scared" on one pass, it will almost surely leave a wake across the path of the helicopter's next pass. After spotting the wake, the helicopter can quickly run down the target. In effect, the team has gotten the advantage of a 700-ft sweep width, without leaving gaps in coverage.

I would recommend two changes in the North Carolina team's final plan. First, their back-and-forth plan calls for the helicopter to make a U-turn every 30 to 60 seconds, which is probably impractical. It would be better to lay the tracks out along the length of the search area, even though the wake would have more time to age between one pass and the next. Also, as noted above, I would try to search the highest-probability area first, even at the expense of an extra transit or two. My recommendation is therefore something on the order of **Figure 3**. In fact, that would be my recommendation whether wakes are detectable or not. I do not know whether this plan is really better than the North Carolina team's plan, or the best of the Drake University team's quadrangular plans.

Some idle questions: If the inference in the problem is correct, won't there be *two* boats in the area? If wakes are detectable, wouldn't the target be better advised to remain stationary?

The North Carolina team poses a question for the helicopter that might be asked of a dog chasing a truck: When you catch it, what are you going to do with it? In fact, there are ways for a well-armed helicopter to apprehend a boat: either lower a boarding party, or escort the boat to port under threat of sinking. Both methods take time, which must be allowed for in planning the search.

The Drake University team is to be congratulated for their simulation program. It is exactly the tool needed for this kind of problem, since it

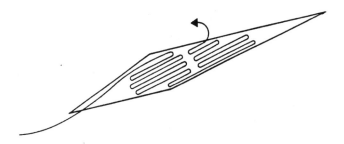

Figure 3. Recommended plan (not to scale).

can be adapted to almost any set of assumptions about searcher and target strategies. To bring such a program to the point of producing usable results in the limited time available is a remarkable accomplishment.

An excellent reference on search theory is Stone [1977].

Acknowledgment

These comments have been prepared with the assistance of Dr. Joseph P. Discenza.

Reference

Stone, Laurence D. 1977. Search theory: A mathematical theory for finding lost objects. *Mathematics Magazine.* 50 (5) (November 1977): 248–256.

1989: The Midge Classification Problem

Two species of midges, Af and Apf, have been identified by biologists Grogan and Wirth [1981] on the basis of antenna and wing length. (See **Figure 1.**) Each of nine Af midges is denoted by "□", and each of six Apf midges is denoted by "○". It is important to be able to classify a specimen as Af or Apf, given the antenna and wing length.

1. Given a midge that you know is species Af or Apf, how would you go about classifying it?

2. Apply your method to three specimens with (antenna, wing) lengths $(1.24, 1.80)$, $(1.28, 1.84)$, $(1.40, 2.04)$.

3. Assume that species Af is a valuable pollinator and species Apf is a carrier of a debilitating disease. Would you modify your classification scheme and if so, how?

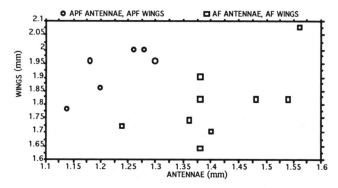

Figure 1. Display of data collected by Grogan and Wirth [1981].

Comments by the Contest Director

The problem was based on Grogan and Wirth [1981, 1285].

Reference

Grogan, William L., Jr., and Willis W. Wirth. 1981. A new American genus of predaceous midges related to *Palpomyia* and *Bezzia* (Diptera: Ceratopogonidae). *Proceedings of the Biological Society of Washington* 94 (4): 1279–1305.

Neural-Network Approach to Classification Problems

Thomas D. Fields
David M. Krasnow
Kevin S. Ruland
Washington University
St. Louis, MO 63130

Advisor: A.F. Kleiner

Summary

We describe a method for solving classification problems using neural networks. We discuss some of the basic mathematical properties of neural networks, their structure, and their training process. We focus on the use of threshold functions in hidden nodes, and their value in classification problems. Issues of uncertainty and stability are also addressed. Finally, an alternative solution method based on neural networks, Learning Vector Quantization, is discussed for comparison.

Introduction to Neural Networks

Neural networks, based on human cognition, attempt to model the learning and decision-making aspects of neurons. Neural networks work by associating concepts and are thus ideal for this problem.

Neural networks were perhaps best described by R. Hecht-Nielsen, one of the early inventors of commercial neurocomputers. He defined a neural network as "a computing system made up of a number of simple, highly interconnected processing elements, which processes information by its dynamic state response to external inputs" [Caudill 1987].

Since neural networks are commonly represented as directed graphs (see **Figures 1** and **2**), the processing elements can be called *nodes* and the interconnections *arcs*. The *value* of a node is a function of the sum of the values of previous nodes multiplied by the weight along the connecting arc.

In fairly straightforward systems such as ours, the nodes are strictly ordered into layers such that a node in layer n transmits or *fires* only to nodes in layer $n + 1$. Our model uses three layers. The first, the *input layer*, receives data from only external sources. The *middle* or *hidden layer* is

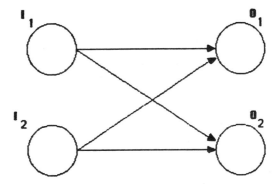

Figure 1. Two-input, two-output neural network.

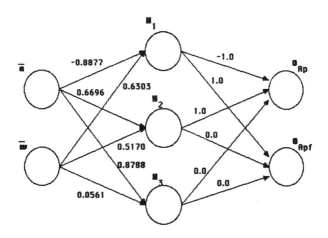

Figure 2. Diagram of a 2–3–2 neural network

stimulated by the input layer, and in turn transmits to the third layer, the *output layer*. The output layer does not transmit to any other nodes; rather, its output is the output of the system.

Given a set of inputs and expected outputs the network is to satisfy, a heuristic is executed that modifies the values of the weights along each arc. This process is known as *training*. After a sufficient number of training runs, the weights on the arcs should approach an equilibrium. At the equilibrium, the interconnections possess the weights that most accurately produce the expected output for the inputs on which the network has been trained. The changes are based on the association between the two nodes on either end. If the training data indicate a strong correlation between the two nodes, the weight will have a large absolute value; if it does not, the weight will stabilize near zero.

This use of association allows networks to classify even unfamiliar input. The importance of associations between the input and output of the network matches its obvious importance in the general classification problem. The biological thought process makes similar use of associations, thereby giving the neural network method and its solution a certain degree of intuitive appeal.

Basic Assumptions

- Species classification of midges is largely dependent on antenna and wing length.

- There is a relatively sharp distinction between species in regard to antenna and wing length.

- The 15 data points are representative of the population.

Design

Structure

In designing the neural network, we found that two input nodes, one each for the antenna and the wing length, yielded the most accurate results. The input can therefore be represented by the vector

$$\vec{I} = (\text{alen}, \text{wlen})$$

where alen and wlen represent antenna length and wing length of a midge. Several other measurements were considered, primarily relating to the angle the input vector or its normalization forms with the axis. Two such measures were the ratio of antenna to wing length and the angle between the vector and the positive horizontal axis. Addition of either of these measures to the model resulted, however, in a degradation of performance. [EDITOR'S NOTE: For space reasons, we omit a table of results.]

Error is calculated as the sum of the absolute value of the differences between the expected output and the actual output. We use this measure several other times to determine appropriate parameter settings.

Our model contains two output nodes, one for each species of midge. Like the input, the output can be shown in vector form, as

$$\vec{O} = (O_{\text{Af}}, O_{\text{Apf}}).$$

Output node values range from 0 to 1. A value of 1 represents near-certainty that the input data came from a member of the corresponding species. A value of 0 indicates that the input data did not come from a member of the corresponding species. An Af midge would have output vector $\vec{O} = (1 \quad 0)$ and an Apf would have output vector $\vec{O} = (0 \quad 1)$. Output values in the middle of the range imply uncertainty.

Two inputs and two outputs provide a total of four interconnections (see **Figure 1**). Although this simple network, commonly called the *perceptron*, has many desirable properties, it also has severe drawbacks. Because of its simplicity, it has a limited ability to generalize when faced with unfamiliar inputs. Further, it is usually necessary to employ many input and output nodes in order to attain even this limited ability.

The alternative that we have taken is to include a hidden layer with at least one hidden node. Adding this layer changes the number of arcs to

$$\#\text{arcs} = (\#\text{input nodes}) \times (\#\text{hidden nodes})$$
$$+ (\#\text{hidden nodes}) \times (\#\text{output nodes}).$$

We tested networks with between two and five hidden nodes. The two- and five-node networks were clearly unacceptable due to their high errors, while the other two performed essentially the same. Although the four-node network had a lower error, we chose the three-node network; we felt that little would be gained in terms of the representational value of the system by the increased complexity of the four-node network. Also, for the four-hidden-node network, the system is underdetermined, so the network essentially reduces to the three-hidden-node situation (see **Appendix B**). [EDITOR'S NOTE: For space reasons, we omit a table of results.]

Procedure

The first procedural decision must be made when the input data are initially entered into the network. The data could be entered as given; but

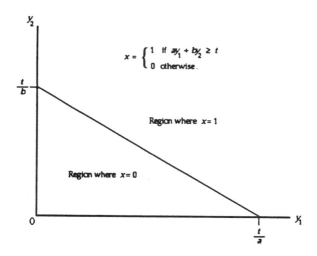

Figure 3. How a threshold function separates a quadrant.

to enhance the stability of our model, we wanted the input node values to be restricted to the range 0 to 1. To achieve this, the stated values, which ranged between 1.1 and 1.6 mm for antenna length and 1.6 and 2.1 mm for wing length, were normalized. This involved simply subtracting 1.1 from each input value.

The procedural decisions regarding the other two types of nodes were considerably more complex. The most important one involved the method of determining the output of a node, given its inputs. There are two basic methods by which this can be accomplished. The first is to use a linear combination of the inputs. Thus

$$\text{output of node} = \sum_{\text{all inputs}} (\text{interconnection weight}) \times (\text{input value})$$

Another method uses a threshold function. For this, the output is binary and depends on the value of the sum of the weighted inputs:

$$\text{output of node} = \begin{cases} 1 & \text{if the sum of weighted inputs} \geq \tau \\ 0 & \text{otherwise,} \end{cases}$$

where τ is the threshold level (see **Figure 3**). In our model, thresholds of the hidden layer nodes play a significant role. Without the threshold function, the network is essentially fitting a continuous linear curve to the points (see **Appendix B**). In our problem, where the expected outputs are binary, this procedure results in fuzzy and imprecise boundaries between the two sets.

planes into regions that have very precise borders. As long as uncertainty (to be discussed later) is considered, accuracy will not suffer.

The most significant aspect of the network is its ability to "learn," or be trained. This is accomplished through the use of a procedure known as *back-propagation of errors* [Rumelhart et al. 1986]. The key to back-propagation is the *delta rule* [Jones et al. 1987], which measures the amount of error and forces appropriate changes in the interconnection weights. Ther are methods for training neural networks; but since back-propagation is capable of assigning deltas to hidden units, which are unable to receive any direct feedback from outside of the network, it is sufficiently powerful.

The first step is to assign a *delta* or error value to each of the output nodes *o*. This is simply the expected output minus the actual output:

$$\delta_o^{\text{out}} = T_o - O_o. \tag{1}$$

Delta terms must also be assigned to the hidden nodes, however. This is more complicated, as it is not known what values the hidden nodes should have. Equation (2) shows that the delta for the hidden node *h* is the sum of the output node deltas (determined by (1)) weighted by the interconnection weights:

$$\delta_h = \sum_i \delta_o^{\text{out}} w_{oh}, \tag{2}$$

where w_{oh} refers to the weight on the arc connecting hidden node *h* and output node *o*.

The term *back-propagation* refers to the feature that the output deltas must be found first. Once the deltas have been determined, the interconnection weights can be modified. The changes in the weights between the input layer and the hidden layer are given by (3) and those between the hidden and output layers by (4):

$$\Delta w_{hi} = \delta_h I_i \ell \tag{3}$$

$$\Delta w_{oh} = \delta_o^{out} H_h \ell, \tag{4}$$

where the learning rate ℓ regulates the speed with which new experiences affect the weights.

It is important to choose an efficient learning rate that leads to a stable network. A high learning rate trains the network very quickly. If it is too high, though, the weights will be extremely unstable; and in the worst case, they will diverge, oscillating around zero with larger and larger magnitudes. A very low learning rate, while avoiding oscillation problems, can mandate an excessive number of training runs.

We tested several values for the learning rate and selected the one with the lowest error after 90 iterations, $\ell = 0.25$ (see **Table 1**).

Table 1.
Learning rate and resulting error.

Learning rate	Error
0.15	1.639
0.20	0.920
0.25	0.000
0.30	0.009
0.35	0.018

Implementation

We implemented the network training in Turbo Pascal on a Zenith 183 IBM-compatible computer. We ran the training process repeatedly, with the changes to the interconnection weights being made as described in (1)–(4).

We chose input data points randomly during the training phase, with equal probabilities assigned to each species rather than to each data point. Although the model converges to the same point regardless of the selection process (so long, of course, as all points are occasionally chosen), our method requires considerably fewer training iterations. We wrote another program on an Apple Macintosh, also in Turbo Pascal, to plot the density graphs of the network's output. Using the arc weights determined on the Zenith, the program calculated output for many points over the entire range. The difference between the two outputs was then translated into a gray scale.

Results

After 90 sweeps of the data (1350 total iterations), the interconnection weights stabilized (see **Figure 2**). The density plot after training is shown in **Figure 4**. The circles represent data for the Af midges, the squares represent data for the Af midges, and the boxed X's are the three test midges. Midges in the solid black region are likely Apf's, those in the white region are likely Af's, and the gray area is a region of uncertainty. All three test points were found to be Af's.

The results for the 4-hidden-node network are virtually identical. Even though it begins with a very different initial position and different random initial weights, after only 90 training runs it converges almost completely to the same solution. [EDITOR'S NOTE: For space reasons, we omit plots for this case.]

Another issue was measuring uncertainty. Some applications might require one to know the certainty of the network's classification decision. In order to deal with this issue, a continuous model (one without thresholds) must be used. Removing the thresholds changes the parameters of the

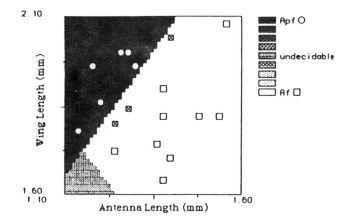

Figure 4. Final solution for the three-hidden-node model, after 1350 training iterations.

model significantly (see **Figure 5**) but should have little impact on the results, except to eliminate their binary nature. Instead, the output takes on real values, mostly between 0 and 1. The output node with the larger value represents the appropriate classification. The difference between the two outputs roughly corresponds to a measure of certainty. **Table 2** illustrates this for the 15 initial data points.

When the three test points were tested, the certainties, as shown in **Table 3**, clearly showed the uncertainty inherent in these boundary points.

Stability Analysis

Because of the complex manner in which the network determines its arc weights, stability is very important. Three distinct types of stability are relevant.

The most basic type of stability is *stability across initial weights*. The randomly-determined starting weights should have little or no effect upon the trained network. However, among the many sets of initial weights that lead to the same equilibrium solution, there is considerable variation in the rate of convergence. Moreover, many sets of initial weights fail to converge to a low-error solution, even after a large number of iterations. The reason may be the existence of local minima in the error-surface, which prevent the finding of a global minimum. The standard remedy is to increase the number of nodes [Rumelhart et al. 1986], but doing this made the situation worse rather than better. For the comparisons in the tables, we selected arbitrarily a starting point that converged to 0 fairly rapidly and held it fixed

Table 2.

Certainty of classification for the 15 data points.

Data point	Output 1 Af	Output 2 Apf	|Difference| ≈ Certainty
1	−0.12	0.57	0.69
2	0.08	0.68	0.62
3	−0.10	1.06	1.16
4	0.15	1.03	0.88
5	0.22	0.99	0.77
6	0.34	0.81	0.47
7	1.16	0.69	0.47
8	0.70	0.45	0.25
9	0.79	0.18	0.61
10	1.16	−0.03	1.19
11	1.38	−0.15	1.53
12	0.80	−0.06	0.86
13	0.38	0.12	0.26
14	0.99	−0.27	1.26
15	0.99	−0.47	1.43

Table 3.

Certainties for the three test points.

Data point	Difference ≈ Certainty
1.24, 1.80	0.09
1.28, 1.84	0.05
1.40, 2.04	0.26

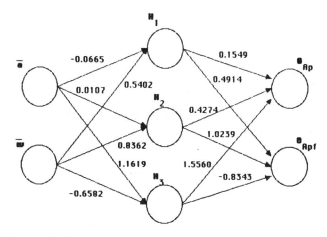

Figure 5. Final solution to the three-hidden-node continuous model.

over the different models and situations being tested.

A second kind of stability is *stability within the training cycle*. A trained network should not oscillate with further training runs. Here our model performed very well. Once trained, the interconnection weights were permanently fixed and no amount of further training appeared to alter them.

The third necessary stability is *stability under perturbations*. Due to the presumably arbitrary selection of the 15 data points, small perturbations in their values should not produce large changes in the model's interconnection weights. We perturbed the data points singly and then all at once. [EDITOR'S NOTE: We omit a table of results.]

In every case, the change to the second tier of weights, between the hidden and output layers, was less than 0.0005; and in many cases it was zero. All of the instability, therefore, is concentrated in the first half of the network. The effect of the latter instability is greatly damped by the threshold function in the hidden layer.

Strengths and Weaknesses

The primary strength of our model is its adaptability. Once the basic network structure is set up, fairly major changes can be made to the problem without involving significant changes in the model. New data points, or even new nodes, can be easily created. Not only can major aspects of the current problem be easily altered, but one can, with few changes, also generalize the model to other classification problems. The same model would be quite effective even with many inputs and many outputs (representing many different classifications, or perhaps represent membership in a multidimensional fuzzy set).

One significant weakness of our model lies in its use of the hidden layer thresholds to produce output that is nearly binary. Most points are either an Af or an Apf midge, and there are very few points about which the network indicates any uncertainty. Ideally, the model would produce probabilistic output, such as a 70% chance of being Af and 30% chance of being Apf. Although the continuous model without thresholds would be an improvement in this respect, its other problems—imprecision and instability—led us to stick with the threshold model. When probabilistic output is required, the two models can be used in conjunction.

References

Caudill, M. 1987. Neural networks primer: Part I. *AI Expert* (December 1987): 47ff.

Jones, W.P., and J. Hoskins. 1987. Back-propagation: A generalized delta learning rule. *Byte* (October 1987): 155–163.

Juell, P.L., K.E. Nygard, and K. Nagesh. Unpublished. Multiple neural networks for selecting a problem solving technique. Department of Computer Science and Operations Research, North Dakota State University.

Kam, M., A. Naim, and K. Atteson. 1988. Smart symmetrized adaptive resonance theoretic model for binary-pattern classification. In *Proceedings of the 1988 Conference on Information Sciences and Systems*. Princeton, NJ.

Kohonen, T. 1988. An introduction to neural computing. *Neural Networks* 1: 11–14.

Levin, S., ed. 1980. *Lecture Notes in Biomathematics: Analysis of Neural Networks*. New York: Springer-Verlag.

Rumelhart, D.E., G.E. Hinton, and R.J. Williams. 1986. Learning representations by back-propagating errors. *Nature* 323: 533–536.

Appendix A
Alternative Method:
Learning Vector Quantization

We also pursued the Learning Vector Quantization (LVQ) approach [Kohonen 1988] for a comparison to the neural network method. LVQ was specifically designed for classification purposes. Because it is derived from, and indeed could be written as, a neural network, it has many similarities in design to the neural network method; but by describing it in algorithmic form, one can completely discard the network structure.

LVQ starts by generating a set of vectors $\vec{m}_i$. These are allocated to the classifications being considered, in the same ratio as the original data vectors. We elected to use 20 vectors, with 8 allocated to Apf and 12 to Af. Each $\vec{m}_i$ is equal to the mean of the data vectors for its class plus a small random vector.

During the training phase, a training element $\vec{x}$ is randomly chosen from the data points. One then determines which $\vec{m}_i$ is most similar $\vec{x}$. Similarity can be defined in many different ways, and the model can adjust to it. We measured similarity with the euclidean norm of the difference, $\|\vec{x}-\vec{m}_i\|$. The $\vec{m}_i$ that is most similar to the training element is then adjusted as follows:

$$\vec{m}_i = \begin{cases} \vec{m}_i + \alpha(k)(\vec{x} - \vec{m}_i) & \text{if } \vec{x} \text{ and } \vec{m}_i \text{ are in the same class} \\ \vec{m}_i - \alpha(k)(\vec{x} - \vec{m}_i) & \text{otherwise,} \end{cases}$$

where $\alpha(k)$ is a monotonically decreasing sequence and k indexes iteration of the process. We used $\alpha(k) = 0.2/k$ and 300 iterations. All other $\vec{m}_i$ remain unchanged (so only one vector is adjusted at each iteration). Through such training, the $\vec{m}_i$ vectors acquire new values but retain their original classifications. The result is a piecewise-linear boundary between the classifications. An arbitrary vector is then assigned to the class of the most similar $\vec{m}_i$, i.e., to the class of the nearest $\vec{m}_i$ neighbor.

For our data, the LVQ boundary between Af and Apf is very close to the line that was produced by the neural network. The three sample data points, which lie very close to the boundary, are classified by LVQ as all Apf—exactly opposite to the neural network model.

We do not feel, however, that this conflict should lead one to conclude that either model is flawed. Fairly small differences in process could easily lead to a slightly different solution boundary. Much of the discrepancy can be easily explained by the clustering nature of the LVQ solution. The $\vec{m}_i$ vectors for the Af class tend to cluster around the class mean of $(1.45, 1.85)$—and hence around the larger concentrations of midges (see **Table A1**)—and ignore outliers, such as the measurement at $(1.24, 1.72)$. The neural network, on the other hand, largely ignores clustering effects, and so does not have this problem.

Table A1.
Vectors generated for LVQ.

Apf	Af
(1.31, 2.02)	(1.41, 1.86)
(1.24, 2.02)	(1.42, 1.81)
(1.27, 2.02)	(1.45, 1.80)
(1.25, 2.00)	(1.42, 1.84)
(1.29, 1.97)	(1.46, 1.86)
(1.22, 2.00)	(1.47, 1.82)
(1.32, 1.98)	(1.48, 1.88)
(1.27, 1.97)	(1.44, 1.88)
	(1.50, 1.84)
	(1.50, 1.80)
	(1.42, 1.83)
	(1.77, 0.75)

Appendix B
Mathematics of Neural Networks

The output value of each node is the sum of the products of the inputs and weights along the arcs.

$$x_i = \sum_j w_{ij} y_j$$

For example, to calculate the output of node O_1 in **Figure 2**, we have

$$o_1 = m'_{11} h_1 + m'_{12} h_2 + m'_{13} h_3$$

and each of the h_i's can be calculated in a similar manner by

$$h_i = m_{i1} i_1 + m_{i2} i_2, \qquad i = 1, 2, 3$$

Combining these gives

$$o_1 = \sum_{i=1}^{3} \sum_{j=1}^{2} m'_{1i} m_{ij} i_j.$$

These formulas for the output of the network in terms of the inputs can be written in matrix shorthand as

$$\vec{O} \equiv \begin{pmatrix} o_1 \\ o_2 \end{pmatrix} = \begin{pmatrix} m'_{11} & m'_{12} & m'_{13} \\ m'_{21} & m'_{22} & m'_{23} \end{pmatrix} \begin{pmatrix} m_{11} & m_{12} \\ m_{21} & m_{22} \\ m_{31} & m_{32} \end{pmatrix} \begin{pmatrix} i_1 \\ i_2 \end{pmatrix} \equiv M_{oh} M_{hi} \vec{I}.$$

Thus, the outputs are simply a linear combination of the inputs. In training a neural net, one tries to find the m_{ij} and m'_{ij} to satisfy all the given training points. In the above example there are 12 variables; therefore, a well-defined system should be trained with 12 given points. If fewer points are given, the linear system is underdetermined, and many solutions exist; if too many points are specified, the system is overdetermined, and solutions may not exist.

Practitioner's Commentary: The Outstanding Midge Classification Papers

Michael B. Richey
Operations Research & Applied Statistics Dept.
George Mason University
Fairfax, VA 22030

As one might imagine, when the judges first saw this problem, we expected to see papers that tried to model the problem by various statistical approaches, including regression analysis, the bivariate normal distribution, linear discriminant functions, and hierarchical clustering. Then judging would be a matter of deciding which teams made the right statistical assumptions and used appropriate statistical techniques.

This assessment was essentially correct; however, a couple of teams opened our eyes to the world of neural networks, particularly the Washington University team.

Conceptually, the problem is fairly straightforward: Given the antenna and wing lengths of a midge, determine whether it should be classified as an Af midge or an Apf midge. However, a number of seemingly useful modeling techniques either contain inappropriate assumptions about the probability distributions of the data, or are unable to model the problem well enough to answer the questions asked.

Linear regression is a well-known statistical model in which one would draw one line for wing lengths and antenna lengths for the Af midges and another line for the Apf midges. Then, for a point in between the lines, one could determine to which species that point corresponds by using the mean and variance of each set of data to determine how far (how many standard deviations) the point is from each line.

However, in building a linear regression model one must assume one variable is the "independent variable" and the other is the "dependent variable," because the model will create different regression lines for these two cases. Since there is no way to view either the antenna length as dependent on the wing length or the reverse, neither regression line would be appropriate. Attempts to fix this problem, such as by computing both lines and averaging them somehow, would fail, because it is not clear how to average the lines; but at least this is better than using only one of the lines.

Some teams tried to collapse the two lengths into one variable, then do analysis on the one variable. The most popular of these was to let

$$\text{ratio} = \frac{\text{wing length}}{\text{antenna length}}$$

be the variable. Unfortunately, the probability distribution of the ratio is very difficult to analyze, which makes this approach difficult to use.

The best classical statistical approach was to assume that the two lengths are distributed according to a joint bivariate normal density. Some teams, including Cal Poly, had the wisdom to test this hypothesis. However, because of the small number of data points, this and many other such hypotheses were difficult to reject. With a probability density function fitted to the data for each species, the lengths for each unknown midge could be plugged in, and the species with the greater density could be chosen as the predicted species. Furthermore, this method is easy to adjust to deal with the two important modeling issues to be discussed next.

One of these issues concerns the relative populations of Af and Apf midges in nature. The two reasonable assumptions are that they occur in equal numbers and that they occur in the 9:6 ratio that appears in the sample data. Since the modelers do not know how the sampling was done, either assumption could be correct. What is important is that the model be capable of dealing with ratios other than 1:1.

The other major issue is how to alter the decision rule to account for the diseased midges. One would like to do this in a controlled manner, so as to estimate the probability of misclassifying either species of midge, and so that only a known, very small percentage of the diseased Apf midges are released back into the wild. This ability to estimate how well the classification has been done is a major strength of the models that possess it, such as the bivariate normal model.

Besides testing the probability distribution assumptions, another model stability issue was to determine how sensitive the model was to the specific input data it received. One approach, used by the Washington University team and others, was to perturb the given data somewhat and see how that affected the model. It generally was found that such perturbations had no significant effect on the models. Another idea was to rebuild the model with 14 data points, then see whether the 15th point would be classified correctly. This sometimes caused the point (1.24, 1.72) to be misclassified when it was the one deleted.

A few more-modern approaches were tried by some of the teams. One of these was to use a hierarchical clustering technique. Unfortunately, this technique could try to misclassify some of the given data, which would leave the modeler in an awkward position. However, hierarchical clustering techniques are quite effective when the data consist only of samples whose categories are unknown.

Another nonstatistical approach was to build a discriminant function (usually linear) that would use some distance measure to determine the boundary between the two species. This and the neural-networks model were the best models for true pessimists, who are reluctant to assume any probability distribution for a set of data. However, the following awkward questions then must be dealt with: How are unequal population sizes (such as 9:6) to be treated? And how can one control the number of diseased midges re-released? A geometric argument may suffice for the first question, but I do not see how to handle the second one.

Other issues were investigated by the Outstanding papers. The Caltech team used a biological argument that the square of the wing length should be proportional to the cube of the antenna length. Then they showed that transforming the data to reflect this had little effect on the model, so they used the untransformed data after all. They also were willing to evaluate their integrals numerically rather than rely on tabulated probability functions, something that many teams apparently did not think of, were unable to do, or were scared off from.

The Caltech team also performed an interesting test on the 9:6 ratio. They assumed that the 15 sample points were taken from a binomial distribution with unknown ratio, rather than assuming the ratio was known to be 9:6. Their analysis showed that essentially the same classification rule was created with and without assuming the 9:6 ratio.

The Cal Poly team brought up the question of whether male and female midges have the same physical characteristics. Since the midges' sexes were not mentioned in the problem description, the team assumed that the midges were all the same sex, or that sex did not matter. If sex mattered, the sample data would need to be split by sex as well as by species.

Also, in their basic model the Cal Poly team pooled the covariance data (of the wing and antenna lengths) for the two species, an approach that works only if the covariances are equal for the two species. Fortunately, they then applied the more complex quadratic classification rule, which does not make this assumption. Essentially the same rule is created by the two methods. If anything, the team was overly critical of the method.

Washington University's neural network was quite sophisticated. Unlike the statistical approaches, which were devised with this sort of problem in mind, neural networks can be applied to a very broad range of applications. One useful aspect of the neural-network approach is that it can model problems with several dimensions or categories. In this sense, applying neural networks to this problem was a bit like using a shotgun to kill a fly. Indeed, it was remarkable that the team was able to apply this general, still-emerging technique to a specific problem in an appropriate manner and within the allotted time.

On the other hand, neural networks have some important disadvantages. For example, although the difference between the two output values of the network gives one some measure of confidence in a decision, this measure has no direct relation to the number of diseased midges that would be misclassified. Furthermore, the Washington University team was unable to adapt their model to assume anything other than a 1:1 ratio of Af to Apf midges in nature.

The other detail that hurt the performance of the neural network was that there were no sample points in the region between the two clumps. Unlike the statistical approaches, which can use probability distributions to draw conclusions about points in the "middle," the only way a neural network can draw good conclusions about such points is to receive some as training points.

Acknowledgment

I would like to acknowledge the assistance M. Habib gave me in preparing this commentary.

References

Cormack, R.M. 1971. A review of classification. *Journal of the Royal Statistical Society Series A* A134: 321–367.

Draper, N.R., and H. Smith. 1966. *Applied Regression Analysis*. New York: Wiley.

Gordon, A.D. 1981. *Classification: Methods for the Exploratory Analysis of Multivariate Data*. New York: Chapman and Hall.

_____. 1987. A review of hierarchical classification. *Journal of the Royal Statistical Society Series A* A150, Part 2: 119–137.

Hartigan, J.A. 1982. Classification. In *Encyclopedia of Statistical Sciences*, vol. 2, edited by Samuel Kotz, Normal L. Johnson, and Campbell B. Read. New York: Wiley.

Hines, W.W., and D.C. Montgomery. 1972. *Probability and Statistics in Engineering and Management Science*. New York: Wiley.

James, Mike. 1985. Classification Algorithms. New York: Wiley.

Panel on Discriminant Analysis, Classification, and Clustering. 1989. Discriminant analysis and clustering. *Statistical Science* 4 (1): 34–69.

Rumelhart, D.E., J.L. McClelland, and the PDP Research Group. 1986. *Parallel Distributed Processing: Explorations in the Microstructure of Cognition*. Vol. 1: *Foundations*. Cambridge, MA: MIT Press.

Tou, J.T., and R.C. Gonzalez. 1974. *Pattern Recognition Principles*. Reading, MA: Addison-Wesley.

About the Author

Michael Richey has been an assistant professor of operations research and applied statistics at George Mason University since 1985. He received a Ph.D. in industrial and systems engineering from Georgia Institute of Technology in 1985, preceded by an M.S. in computer science from the University of Missouri—Rolla in 1981.

His primary research interest is in graph algorithms; however, he also has studied problems in facility location, scheduling, bin packing, and parallel algorithms. He was a Summer Faculty Fellow at NASA Goddard Space Flight Center, and he recently devised a clustering method to help the Defense Communications Agency decrease the size of its network design problems. Another favorite problem of Dr. Richey's has been the application of operations research and statistical methods to the design of the National Football League's schedule.

His favorite teaching topics are Wagner's puny stagecoach problem for dynamic programming, and teaching distributed processing by having each student act like one of the processors. Prof. Richey was a judge for the Midge Identification problem.

Postscript

In 1980, Brazil's Amazon valley saw its worst outbreak of a recurring viral disease called Oropouche. While not fatal, the disease debilitates its victims for several days with severe nausea, muscle pains and in 5 to 10 per cent of cases progresses to an inflammation of tissues around the brain. Health officials have linked the 1980 outbreak and other Oropouche epidemics to huge piles of cacao husks that have accumulated as the crop's cultivation there has increased. Filled with rainwater, the husks serve as breeding grounds for tiny insects, called midges, that spread the disease.

—R. Weiss, Hitching a ride with imported insects,
Science News 136 (23 September 1989): 202.

1990: The Brain-Drug Problem

Researchers on brain disorders test the effects of the new medical drugs—for example, dopamine against Parkinson's disease—with intracerebral injections. To this end, they must estimate the size and the shape of the spatial distribution of the drug after the injection, in order to estimate accurately the region of the brain that the drug has affected.

The research data consist of the measurements of the amounts of drug in each of 50 cylindrical tissue samples (see **Figure 1** and **Table 1**). Each cylinder has length 0.76 mm and diameter 0.66 mm. The centers of the parallel cylinders lie on a grid with mesh 1 mm × 0.76 mm × 1 mm, so that the cylinders touch one another on their circular bases but not along their sides, as shown in the accompanying figure. The injection was made near the center of the cylinder with the highest scintillation count. Naturally, one expects that there is drug also between the cylinders and outside the region covered by the samples.

Estimate the distribution in the region affected by the drug.

One unit represents a scintillation count, or 4.753×10^{-13} mole of dopamine. For example, the table shows that the middle rear cylinder contains 28,353 units.

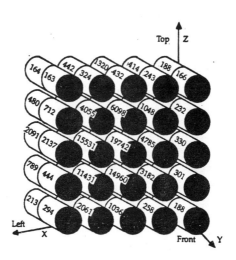

Figure 1. Orientation of the cylinders of tissue.

Table 1.

Amounts of drug in each of 50 cylindrical tissue samples.

Rear vertical section

164	442	1320	414	188
480	7022	14411	5158	352
2091	23027	28353	13138	681
789	21260	20921	11731	727
213	1303	3765	1715	453

Front vertical section

163	324	432	243	166
712	4055	6098	1048	232
2137	15531	19742	4785	330
444	11431	14960	3182	301
294	2061	1036	258	188

Comments by the Contest Director

This problem was contributed by Yves Nievergelt (Mathematics Dept., Eastern Washington University, Cheney, WA). His commentary (p. 135) explains the origin and motivation for the problem. The only adaptation from the applied situation to the MCM problem statement was a slight simplification of the real-world data. The Outstanding papers provided impetus for another study of the data for Seattle's Regional Primate Research Center in the autumn of 1993.

Error-Function Diffusion: A Dopamine–Fick's Model

Michael J. Kelleher
David Renshaw
Athan Spiros
California Polytechnic State University
San Luis Obispo, CA 93407

Advisor: T. O'Neil

Summary

Estimation and modeling of drug diffusion in tissue is necessary for successful combatting of many brain disorders. An obvious method for determining the size and shape of the region of the brain affected by a drug consists of actually removing tissue and measuring the quantity of drug in certain core samples. We were given measurements of 50 core samples spread in three dimensions around the point of injection, and were asked to estimate the size and shape of the region affected by the drug. We treated the data from the core samples as densities rather than scintillation counts by simply dividing the scintillation count by a unit volume of one core. Continuity of the dispersion allowed us to assume that representing the data by point densities within each core is valid.

Research on diffusion in tissue suggested two types of models for estimating both size and shape: an exponential function, and a distribution based on Fick's law. The latter offers a general description for similar diffusion problems, and the application of appropriate boundary conditions yielded a superior model for the dispersion of dopamine based on the standard error function (erf). Subsequent modifications yielded a still better fit. Our final model was the function

$$C(x, y, z) = 51103 - 51319 \text{ erf } [0.6(x + 0.3)^2 + (y + 0.1)^2 + (z + 0.4)^2],$$

a function of position yielding the dopamine concentration at the point (x, y, z).

We evaluated the quality of our model by measuring the average absolute residual between our estimator and the data points. We also compared the residuals with those from an exponential model, which was fairly accurate, and found that the erf model was better still.

We were tempted to make assumptions about our data based on the shape of computed level curves and surfaces. However, the arguments for many of these were so weak as to merit no investigation into further modifications of the model.

Assumptions

We made the following assumptions:

1. We will not be asked to describe past or future distributions.

2. Spread of the drug is due solely to diffusion and to a monodirectional current.

3. The tissue is homogeneous, containing no preferred routes of transfer for the dopamine.

4. There are no boundary surfaces.

5. The density distribution is continuous with respect to position.

6. The initial injection of dopamine occurred exactly in the center of the middle rear cylinder and was instantaneous.

7. Scintillation counts are converted to amounts per unit volume and each density is assumed to be at the center of the measured cylinder.

8. The absorption of the drug does not affect its dispersion pattern.

9. All 50 core samples were removed simultaneously.

Justification of Assumptions

1. **One point in time:** Since no information was given on how much time has elapsed since the injection, the rate at which the distribution is changing cannot be determined.

2. **Reasons for dispersion:** Movement of extracellular substances does occur in the cerebrospinal fluid bathing the brain [Marsden 1984, 116]. Literature suggests that both constant and transient currents exist within this fluid. However, the nonconstant flows are small in comparison to the steady currents. From graphs of the data, we see that the distribution is fairly constant with respect to distance from the highest concentrations, indicating that no major distortions exist which could be expected from nonsimple currents.

   The literature also states that "Major transport of oxygen, carbon dioxide, glucose, and metabolites is due not to net movement of fluid but to simple diffusion across the capillary walls and through the interstitial fluid"; so diffusion can be expected to account for the basic distribution, allowing a current to make secondary contributions [Bradbury 1979, 1, 27].

3. **Tissue homogeneity:** Brain tissue is a collection of nerve cells that are macroscopically similar [Gardner 1975, 136]. The cerebrospinal fluid bathing the tissue is also of a consistent makeup. The radial symmetry observed earlier suggests that there are no preferred routes for the dopamine.

4. **Boundary surfaces:** We have no knowledge of the environment outside of the samples taken. So we have no reason to assume that there are any membranes or other disruptive boundaries within the affected region of the brain which might influence the distribution of the drug.

5. **Continuity of dispersion:** Natural dispersion processes tend to be continuous, and there is no reason to assume that discontinuities exist.

6. **Point of injection:** The injection was made near the center of the cylinder with the highest concentration, and there is no reason to assume that it was not made at the exact center. We regard the injection as a point-source impulse and not as a directed velocity with an interval of injection.

7. **Density measure:** Some measure of density distribution is needed. Since the dispersion within each cylinder is unknown, its center is a logical choice to place the data. Also, the amount of the dopamine within the cylinder, divided by its volume, is the average density within this structure. Since the distribution is already assumed to be continuous, the intermediate value theorem ensures that there is a point within the cylinder which has this density. The true location for a point with this density cannot vary farther from the center than the boundaries of the cylinder itself.

8. **Drug absorption:** No information is given about the absorption rates of the drug into the tissue, so no meaningful influences of nonlinear absorption can be inferred.

9. **Simultaneous core removal:** If this assumption is not made, then data on when each core was removed is essential to any model.

Analysis of the Problem

As is often the case, a good first step with this problem is to draw a picture. While it is difficult to draw three dimensions in space and a fourth dimension for dopamine concentration, we can easily separate the data into two groups, the front and the back, and then draw two sets of level curves.

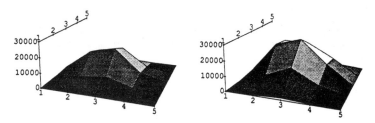

Figure 1. X-Y position versus concentrations, for front (left) and back (right).

Doing so gives a good idea of the "shape" of the data. In **Figure 1**, we provide one view of each row of tissue samples.

We see from the contour maps in **Figure 2** that the data are symmetric about some center. Since we made the assumption that the only way the drug could spread was through diffusion and outside forces, these graphs lead us to believe that the dominant dispersion was due to diffusion.

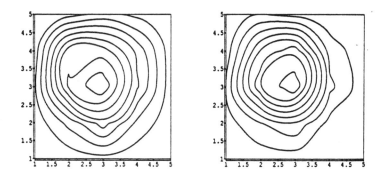

Figure 2. Contour view of sample data, for front (left) and back (right).

The data in the figures are treated as ordered triplets (x, y, c) for the convenience of being able to create a graphical representation, where c is the concentration. In the following estimator models, however, the data are treated as ordered quadruples (x, y, z, c), as there are three spatial dimensions in the data. Our coordinate system is centered at the injection point, placing $(0, 0, 0)$ in the center of the middle back cylinder. Our x-axis extends negatively to the left, the y-axis extends positively from rear to front, and the z-axis extends positively toward the top.

The Exponential Model

Our first model of the shape and size of the region of diffusion is based on an exponential estimator. Exponential models are commonly used to describe diffusion in a homogeneous medium [Twiner 1962, 40]. The exponential estimator we use is a function of the distance from the point of highest concentration. If diffusion is the only force, then this point would be the point of injection, $(0, 0, 0)$. Taking into account other forces, such as currents, would change the situation. We use the exponential function

$$C(x, y, z) = B \exp \left(A \text{ dist } (x, y, z) \right),$$

where C is the concentration at the point (x, y, z) and dist (x, y, z) is the Euclidean distance from the highest point of concentration to the point (x, y, z). We transform the data by taking the natural logarithm of the concentrations and find the linear estimator

$$\ln \left(C(x, y, z) \right) = 11.81 - 2.30 \text{ dist } (x, y, z).$$

Regression analysis of this fit to the data is surprisingly encouraging. There are only two points indicated by our software as outliers (lying more than two standard residuals away), and the R^2 coefficient is 83.2%, more than high enough to indicate good correlation.

One of the outlier points is the point with the highest concentration. The cause for this is obvious, as an exponential function grows very rapidly at its point of symmetry, far more rapidly than is useful for our model. While this feature may seem a necessary evil of the model, we prefer to look for a model that more closely resembles the data.

Our first modification is to assume that the dopamine experienced the effects of a current that shifted the point of highest concentration to a new center. With the center at $(-0.3, -0.1, -0.4)$, the only change in our model of a concentration function is the distance function. The new distance function measures the Euclidean distance from the point (x, y, z) to the new proposed center. Using linear regression on the transformed variables, we obtain

$$\ln \left(C(x, y, z) \right) = 11.85 - 2.24 \text{ dist } (x, y, z).$$

The R^2 coefficient is 88.1%, indicating a better fit than the original model. Another indication of better fit is the average absolute residual. In the previous model it was 3.86×10^3; in the new model it is 1.84×10^3.

Fick's Law and the erf Function

In our efforts to find a better model for the diffusion, we came across Fick's law, a common model for diffusion. The basic form of Fick's Law is

$$(-AD) \frac{\partial C}{\partial R} = J,$$

with

A	the cross-sectional area for flux,
D	the diffusion coefficient for binary mixture,
R	the distance of point from center of diffusion,
$C(R, T)$	the concentration as a function of R at time T, and
J	the flux of diffusing material.

E.L. Cussler [1976, 19] gives the following technique for using this equation to solve for the concentration levels. First, the mass balance equation

$$-\frac{\partial C}{\partial T} = \frac{\partial J}{\partial R} + CV$$

is used, with

T	time and
V	the velocity of fluid in which diffusion occurs.

Fick's law indicates what flux, or movement of the diffusing material, is generated by the imbalance of concentrations. The mass balance equation determines how the concentration at each point is affected by amounts of the substance diffusing through it. For now, we assume that $V = 0$; that is, we ignore any mass movement of the drug because of flows.

Combining the two equations yields the partial differential equation

$$\frac{\partial C}{\partial T} = AD\frac{\partial^2 C}{\partial R^2}.$$

With the substitution

$$Y = \frac{R}{\sqrt{4ADT}},$$

so that

$$Y = FR \quad \text{with} \quad F = \frac{1}{\sqrt{4ADT}},$$

the equation takes the form

$$\frac{d^2 C}{dY^2} + 2Y\frac{dC}{dY} = 0.$$

The grouped constant F accounts for the unknown values of D and T, both of which are constant, since the tissue is assumed to be homogeneous and the concentration function describes a particular time. The general solution to this equation is

$$C = K + B \int_0^Y e^{-x^2} dx.$$

Using the boundary conditions

$$C(\infty) = 0 \text{ and } C(0) = C_0,$$

we solve for the related constants K and B:

$$C(0) = K + B \int_0^0 e^{-x^2} dx = C_0, \text{ so } K = C_0;$$

and

$$C(\infty) = K + B \int_0^\infty e^{-x^2} dx = B\left(\frac{\sqrt{\pi}}{2}\right) + K = 0,$$

so

$$B = -\frac{(2K)}{\sqrt{\pi}} = \frac{(-2C_0)}{\sqrt{\pi}}.$$

Therefore, the concentration as a function of distance from the center of the diffusion process is

$$C(Y) = C_0 + \frac{(-2C_0)}{\sqrt{\pi}} \int_0^Y e^{-x^2} dx = C - C_0 \operatorname{erf}(Y),$$

where $Y = FR$ and erf is the standard normal error function

$$\operatorname{erf}(x) = \frac{2}{\sqrt{\pi}} \int_0^\infty e^{-x^2} dx = \Phi(x) - \frac{1}{2},$$

where Φ is the cumulative distribution function of a standard normal random variable (with mean 0, variance 1).

In order to determine the value of F, the center of the distribution is assumed to be at the origin and C_0 is assumed to be the round number 30,000, which is slightly higher than the largest concentration observed. The values of F needed to generate a few of the raw data points are then determined from their distances from the center. For example, using the point with scintillation count 3182, we have $R = 1.61$ and

$$30,000 - 30,000 \operatorname{erf}(FR) = 3,182,$$

so $F = 0.7135$ from tables for the error function [Beyer 1986, 527].

The values found indicate that F lies between 0.3 and 0.8. We next do least-squares fits for the coefficients for different F values within this range, varying the F value until the sum of the absolute values of the residuals is minimized. The best average absolute residual for the 50 data points is 2,130, with corresponding function $36,334 - 37,258 \operatorname{erf}(0.60R)$.

We notice that the coefficients for the constant and erf term are nearly equal, just as predicted by the general solution. This agreement can be taken as evidence that this curve fits the results predicted by theory. The

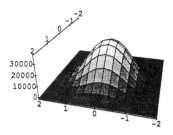

Figure 3. The erf function model for the back row.

concentrations produced by this equation in the plane containing the centers of the rear cylinders look like **Figure 3**.

If any currents exist within the cerebrospinal fluid where the drug was injected, the result would be a translation of the distribution. The concentrations would move evenly if the flow is constant and not experience any distortions as a result of the current.

Plots of the data suggest that the center of the distribution is slightly to the left and down from the middle of the tissue sample. The decrease of the values from front to back indicates also that the center is behind the origin.

By perturbing the position of the center and the value of F, a new fit can be found with residuals smaller than for other positions in the region. With the center at $(-0.3, -0.1, -0.4)$, the residuals are minimized over F at $F = 0.69$, with average absolute residual of 1270 and corresponding function $51,103 - 51,319 \, \text{erf}\,(0.69R)$. Once again, we notice that the values of the constant and erf terms are nearly the same, as they should be to ensure the boundary conditions at infinity.

A further method for analyzing the fit to the data is to pick a core and compute the corresponding level curve, to see if the level curve intersects the core. If it does, the model correctly predicts the measured density for some point in the cylinder. For our fit, not only does every level curve intersect the corresponding cylinder, but each comes closer to the center of the cylinder than to any extreme edge.

We show in **Figure 4** the concentration plots for the planes passing through the centers of the front and back cylinders.

Results

Using the last erf estimate, we are able to describe easily the shape of the dispersion. The region with dopamine in it has to be a sphere around the center chosen, because of the symmetry of the function we use. The only question is how large its radius is. We may assume, for example, that the drug is not effective at concentrations below 50 scintillation counts per

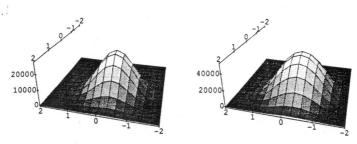

Figure 4. Modified erf model for front (left) and back (right) rows.

unit volume, and proceed to solve the equation at this bound, arriving at a distance of 2.86 mm. Therefore, the effective drug is in the sphere centered at $(-0.3, -0.1, -0.4)$ with radius 2.86 mm. Beyond 2.94 mm, the model predicts no dopamine. It is reassuring to have this touch of realism in our model, since many other models predict traces of dopamine out to infinity.

One might be tempted to try to find a distorted shape to the dispersion; but without strong arguments based on either the specific host tissue or the flow of cerebral fluid, all such arguments rest too heavily on shaky assumptions. A spherical dispersion pattern seems unsatisfyingly simple; but without other information, there is no reason to apply any nonspherical model, especially since the erf model fits so well.

Strengths and Weaknesses

The greatest strength of this model is simply that the estimator function fits the sample data so well. The relatively low average value for the absolute residuals shows that the function prescribes concentrations that are in line with the measured data. Furthermore, the mathematics of the model is deeply entrenched in appropriate physical models of diffusion.

The error function used also allows for extrapolation of the estimator to determine concentrations outside the sample area without losing precision. Other methods that depend solely on connecting the data points, such as splining or brute-force numerical methods, might generate functions that coincidentally fit the data points; but those functions cannot be used effectively for estimating the overall shape of the affected region.

Our solution allows for an extension of the model to a time-dependent solution. If data were available for the dispersion of the drug at other times, then the value of D, the diffusion constant in our equations, might be determined. This would yield a further variation that would be useful in producing estimators for all time values.

Since the method predicts concentrations that constantly decay as a function of distance, level surfaces can be determined for any value of the drug concentration. This is a simple way of determining drug threshold levels, which may be necessary in dosage sizes and redosage times.

The model suffers some from its dependence on distance from the point of highest concentration. Quite possibly our minimum average residual is not the true minimum, since there is uncertainty about the point of highest concentration. Fortunately, we know that the point we use as center is close to the true center, thus guaranteeing a close approximation to the true minimum average residual. Our iterative method of numerical solution is straightforward but imprecise. An analytic solution may exist, but we feel having one could improve the model only slightly.

The plots of the data (without splining or other interpolating functions) may indicate some asymmetry that is not present in our model. Distortions may in fact exist in the data, but modeling them without understanding why they occur is a waste of resources.

Possible improvements of this model abound. For example, we assumed, based on the continuous nature of the diffusion function, that we could treat the scintillation counts as point densities in the center of corresponding cylinders. Given sufficient time, one could transform the data into probability density functions centered in the cylinder and assess the probabilities that the densities, in fact, were at the center of each cylinder.

Further extensions could account for possible asymmetries in the data, based on irregularities in the neural tissue, initial needle velocity of the injection, nonsimultaneous core removal, or other factors. Also, it may be possible to modify the model to account for nonlinear absorption rates, disruptions due to transport cavity irregularities, or dopamine bonding, if there are strong arguments for the presence of such phenomena.

References

Bradbury, Michael. 1979. *The Concept of a Blood-Brain Barrier*. Chichester, UK: Wiley.

Beyer, W.H. 1986. *Standard Mathematical Tables*. 25th ed. Akron, OH: Chemical Rubber Company Press.

Cussler, E.L. 1976. *Multicomponent Diffusion*. New York: Elsevier.

Gardner, Ernest M.D. 1975. *Fundamentals of Neurology*. Philadelphia, PA: Saunders.

Marsden, C.A. 1984. *Measurement of Neurotransmitter Release In Vivo*. Chichester, UK: Wiley.

Twiner, Sidney B. 1962. *Diffusion and Membrane Technology*. New York: Reinhold.

Judge's Commentary:
The Outstanding Brain-Drug Papers

Mario Martelli
Mathematics Dept.
California State University
Fullerton, CA 92634

The contest problem is of significant importance in the treatment of Parkinson's disease and similar disorders. Medical practitioners rely on experience and common sense in deciding on dosage, frequency, and location of the injections.

The approaches submitted by the 73 teams that chose this problem ranged from too simple to quite elaborate. At the one end of the spectrum, we found papers in which a few parameter-dependent functions were proposed as possible descriptions of the distribution of dopamine and the most suitable was selected on the basis of a statistical analysis. At the other end, we found papers in which the temporal and spatial distribution of absorbed and free dopamine were modeled separately as diffusion processes governed by suitable partial differential equations; the two processes were connected by the assumption that the time derivative of the concentration of absorbed dopamine is proportional to the concentration of the free dopamine.

The judges felt that a successful model should strike a compromise between the complexity of the process and an objective representation of the features essential to its useful analysis.

The winning teams were somewhere in the middle of the spectrum. Both provided a list of reasonable assumptions and explained their importance and relevance to the selected model. In the stage of model development, both teams used a diffusion process governed by a partial differential equation in the concentration. The California Polytechnic team assumed spherical symmetry and used Fick's law, while the Humboldt State team did not assume spherical symmetry but regarded the process as having an instantaneous point source at the start of the process. Both approaches probably oversimplify important features of the process.

In particular, the point-source assumption may be regarded as too coarse an approximation to the reality of an injection. Other teams, in fact, attributed to the dopamine an initial velocity, either radial or in a certain direction. To obtain a reliable estimate of the velocity, often a local hospital was called to get information on the size of the needle, the amount of dopamine normally administered, and the duration of the injection. The judges were impressed by these efforts; but they decided that for the problem at hand, the point-source assumption is acceptable, at least as a first approximation.

Spherical symmetry, too, may be flawed; and several teams dismissed it as inappropriate. The California Polytechnic team argued that modeling distortions without understanding why they occur is a waste of resources, although they mentioned the inclusion of possible asymmetries as one of the ways of improving their model. The inclusion is straightforward, and the judges felt that it constitutes a better strategy.

Making a reasonable decision about the determinant features of the process is not easy, and the final choice may be a matter of debate. For example, the background dopamine level was neglected by all teams, although some of them mentioned that they were not sure if this assumption was appropriate. Also, some teams mentioned possible transport currents, and others considered them negligible with respect to diffusion. Again, the judges felt that both these features could be neglected in an initial model.

We hope that use of this problem in the MCM will help encourage the medical community to consider a modeling approach in efforts to improve the effectiveness of dopamine therapy for Parkinson's disease.

About the Author

Mario Martelli has been professor of mathematics at California State University, Fullerton, since 1986. He divides his professional time among teaching a variety of courses (including a new course on Discrete Dynamical Systems and Chaos), doing research in applied mathematics, and supervising teams of Harvey Mudd College students working on collaboration projects with local industries. The most recent efforts of his teams include the study of neural networks as image classifiers (for General Dynamics) and the analysis of electron density and energy in the lower Van Allen radiation belt (for McDonnell Douglas).

Prior to coming to California, he was professor of mathematics at Bryn Mawr College in Pennsylvania and at the University of Florence, Italy, where he received his Ph.D. in 1966. He has been visiting professor at the the universities of Warwick, California–Berkeley, California–Davis, Colorado, and Bonn. His primary research interests are in dynamical systems.

Prof. Martelli was an associate judge for the Brain-Drug Problem.

Author's Commentary:
The Outstanding Brain-Drug Papers

Yves Nievergelt
Dept. of Mathematics
Eastern Washington University
Cheney, Washington 99004

The problem reviewed here—the estimation of the spatial distribution of a drug injected in brain tissue—has a story that demonstrates that outstanding undergraduates can contribute to applied mathematical research.

In the autumn of 1982, Paula Altschul, a former precalculus student, introduced me to Mark F. Dubach, a friend of hers who was completing a dissertation in physical anthropology on the effect of intracerebral dopamine injections. He needed a mathematical estimate to convince both himself and his dissertation committee of the scientific validity of his assessment of the region of the brain affected by the drug.

At the time, even at the University of Washington's prestigious health sciences research complex, the computational equipment available limited the testing of a mathematical model: The IMS Associates 8080 microcomputer offered only 56KB of internal memory and supported only a rudimentary version of BASIC, which ran overnight to perform a nonlinear regression.

To accommodate such equipment, we first assumed a distribution that uniformly filled an ellipsoid (to be estimated through regression) and vanished outside. We grouped each row of five cylindrical punches into a parallelepiped, intersecting the ellipsoid in a region shaped like a candy bar near the center and like a piece of a cheese-wheel near the boundary. This crude mathematical model had an analytical expression involving only inverse tangents and algebra, which the equipment could handle. A few months later, the results became a part of Dr. Dubach's dissertation [1983, 356], while the details of the integral calculus appeared in Nievergelt [1984].

Several years later, but only a couple of blocks away from Dr. Dubach's laboratory, we finally obtained the time, money, and permission to use the University's CDC Cyber 180/855 mainframe computer and the FORTRAN software routines of the International Mathematical and Statistical Library (IMSL). That system took about a minute to fit a homogeneous anisotropic diffusion and absorption model [Nievergelt 1990], which resembles the two Outstanding MCM papers.

The paper from California Polytechnic State University contains interesting simplifying assumptions. For instance, the replacement of three-dimensional integration by a three-dimensional mean-value theorem constitutes an insightful application of advanced calculus in a situation limited by time. As another example, the assumptions of a constant cross-sectional flux area and of a single space-time variable lead to a distribution modeled by the "error function," which closely fits the data but lacks a derivative at the origin. The lack of a derivative suggests that the drug spread only during the injection; otherwise, the elliptic diffusion equation would have yielded an analytic solution [Hörmander, 1963, 101]. Though Hörmander's work still remains beyond the reach of most undergraduates (at least in the U.S.), its applicability may convince students that abstract theory can provide important insight into practical phenomena.

The team from Cal Poly also mentions the drawbacks of using splines to interpolate the data. Grace Wahba of the Statistics Dept. of the University of Wisconsin–Madison notes that an alternative approach would be to use smoothing splines, which do not fit the data exactly, but smooth it [Wahba 1990].

The paper from Humboldt State University, in contrast, derives from the diffusion equation an analytic, anisotropic, Gaussian distribution. Their original FORTAN programs to fit the distribution to the data is less fancy than the use of Mathematica and may appear to be reinventing the wheel; but the design of programs for integration, for optimization, and for solving equations also demonstrates strong resourcefulness to solve a problem from scratch. This paper, too, made an interesting simplifying assumption, that the principal directions of the diffusion lie in the directions of the coordinate axes. The assumption leads to a diagonal diffusion tensor, instead of a more general quadratic form represented by a positive definite (symmetric) matrix (the last topic may have disappeared from many linear algebra curricula).

The two Outstanding papers suggest several conclusions:

- There are outstanding undergraduates capable of applying abstract mathematics and sophisticated computational systems to a concrete problem, at a level at which they could assist researchers.

- They are able and willing to learn additional theory, which would considerably narrow the gaps between the current undergraduate mathematics curriculum and, on the one hand, real applications of mathematics, and, on the other, the allegedly better preparation of foreign students.

- The Mathematical Contest in Modeling is effective in providing an incentive for students to learn mathematics and in providing the community with an objective measure of what undergraduates can *do*.

References

Dubach, Mark F. 1983. Intracerebral dopamine injections: Effects on the unrestrained behavior of long-tailed macaques. Ph.D. diss., University of Washington.

Hörmander, Lars. 1963. *Linear Partial Differential Operators*. Berlin: Springer-Verlag.

Nievergelt, Yves. 1984. A volume integral for medical research. *International Journal of Mathematics Education in Science and Technology* 15 (5): 649–652.

Nievergelt, Yves. 1990. Fitting density functions and diffusion tensors to three-dimensional drug transport within brain tissue. To appear in *Biometrics* 46 (4).

Wahba, Grace. 1990. *Spline Models for Observational Data*. Philadelphia, PA: SIAM.

About the Author

Yves Nievergelt graduated in mathematics from the École Polytechnique Fédérale de Lausanne (Switzerland) in 1976, with concentrations in functional and numerical analysis of PDEs. He obtained a Ph.D. from the University of Washington in 1984, with a dissertation in several complex variables under the guidance of James R. King. He now teaches complex and numerical analysis at Eastern Washington University.

Prof. Nievergelt is an associate editor of *The UMAP Journal*. He is the author of several UMAP Modules, a bibliography of case studies of applications of lower-division mathematics (*The UMAP Journal* 6 (2) (1985): 37–56) (in which the Brain-Drug Problem was discussed explicitly), and *Mathematics in Business Administration* (Irwin, 1989).

1991: The Steiner Tree Problem

The cost for a communication line between two stations is proportional to the length of the line. The cost for conventional minimal spanning trees of a set of stations can often be cut by introducing "phantom" stations and then constructing a new *Steiner tree*. This device allows costs to be cut by up to 13.4% (= $1 - \sqrt{3}/2$). Moreover, a network with n stations never requires more than $n - 2$ points to construct the cheapest Steiner tree. Two simple cases are shown in **Figure 1**.

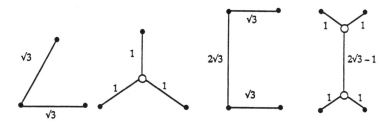

Figure 1. Two simple cases of forming the shortest Steiner tree for a network.

For local networks, it often is necessary to use rectilinear or "checkerboard" distances, instead of straight Euclidean lines. Distances in this metric are computed as shown in **Figure 2**.

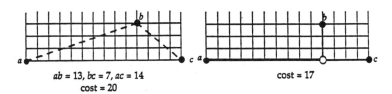

$ab = 13, bc = 7, ac = 14$
cost = 20

cost = 17

Figure 2. Comparison of distances using straight Euclidean line distances ($ab = 13$, $bc = 7$, $ac = 14$; cost = 20) vs. using rectilinear distances (cost = 17).

Suppose you wish to design a minimum cost spanning tree for a local network with 9 stations. Their rectangular coordinates are:

$$a(0, 15), \quad b(5, 20), \quad c(16, 24), \quad d(20, 20), \quad e(33, 25),$$

$$f(23, 11), \quad g(35, 7), \quad h(25, 0), \quad i(10, 3).$$

You are restricted to using rectilinear lines. Moreover, all "phantom" stations must be located at lattice points (i.e., the coordinates must be integers). The cost for each line is its length.

1. Find a minimal cost tree for the network.

2. Suppose each station has a cost $d^{3/2}w$, where d = degree of the station. If $w = 1.2$, find a minimal cost tree.

3. Try to generalize this problem.

Comments by the Contest Director

I (Ben Fusaro) contributed this problem; I got the idea from an article by Barry Cipra [1991].

Reference

Cipra, Barry A. 1991. Euclidean geometry alive and well in the computer age. *SIAM News* 24(1) (January 1991): 1, 16–17, 19.

Finding Optimal Steiner Trees

Zvi Margaliot
Alex Pruss
Patrick Surry
University of Western Ontario
London, Ontario
Canada N6A 5B9

Advisor: H. Rasmussen

Introduction

We find a minimum rectilinear Steiner tree for the network given. We prove or cite some important results concerning minimum spanning trees and Steiner trees, in particular:

- There is a lower bound on the tree length (**Theorem 1** below).

- There is a minimal Steiner tree that spans the network (**Theorem 2**).

- We can find a reduced set of lattice positions for phantom points; in our case, we need to consider only 31 positions out of a possible 936 lattice points (**Theorems 3** and **5**, and **Figure 1**).

- A network of n stations will require no more than $n - 2$ phantom stations (**Theorem 4**).

- The general problem is NP-complete [Chung and Whang 1979].

- We use an algorithm to construct a minimum spanning tree from a given set of fixed and phantom points (see Minieka [1978]).

Using these results, we solve the problem with two basic general approaches:

- **Brute Force:** Our computer program cycled through all possible combinations of phantom point locations, coming up with the absolute optimal spanning tree in about 3.6 million iterations (17 hours of computing time).

- **Simulated Annealing:** A more elegant and efficient way of solving NP-complete problems, the simulated annealing program proved much faster (about 1.5 min for Part 1, 3 min for Part 2) than brute force and consistently (100 times out of 100 tries) converged to an optimal path found by the brute-force algorithm.

Both programs are completely general, in that any fixed-station network and any cost parameters can be used.

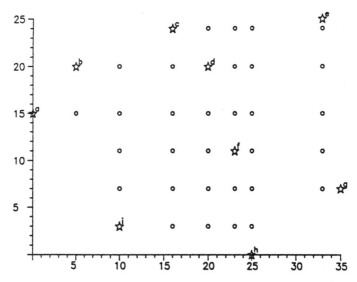

Figure 1. The reduced number of 31 positions that need to be considered for phantom stations. The original 9 points are denoted by lettered stars.

Results

- **Part 1:** We found five different optimal trees, each with either four or five phantom stations and having length exactly 94 (**Figure 2**).

- **Part 2:** We found two trees that are probably optimal, each with two phantom stations of degree three and each costing 135.89 (**Figure 3**).

General Discussion and Theory

Given a set of fixed stations in the taxicab metric $d(x,y) = \|x - y\|_1 = |x_1 - y_1| + |x_2 - y_2|$ on the lattice ^{2}, we are asked to find a minimum spanning graph to network all the stations together. We may add "phantom" stations in order to create a Steiner-type tree. As specified in the problem, the tree branches ("edges") and all station locations must lie along the lattice.

In finding a solution, we have three degrees of freedom:

- how many phantom stations are to be included,

- where each phantom station will be located, and

- how we construct a spanning graph of minimum cost.

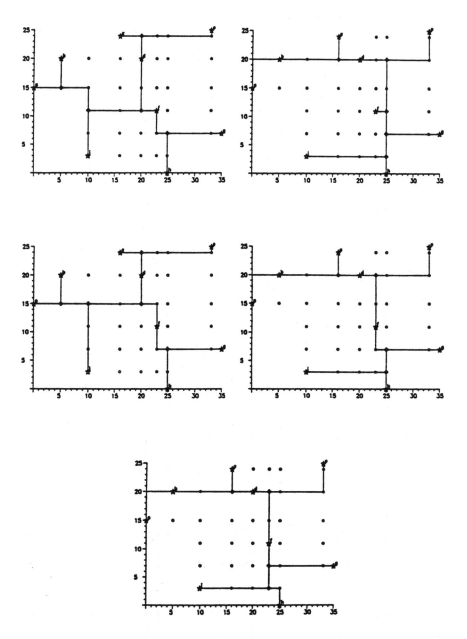

Figure 2. The five optimal solutions to Part 1, each with total edge length of 94. Stars indicate the original fixed points, and crosses indicate added phantom points.

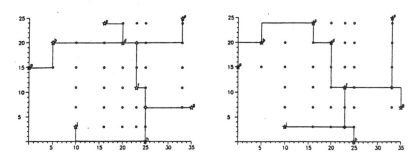

Figure 3. Two probably optimal solutions to Part 2, each with total cost 135.89.

Preliminary Results

- There is an optimal configuration for the given problem. Without one, we could be wasting our time. The result follows from **Theorem 2** below.

- Of the possible phantom-point locations—in our case, the finite set of lattice points within and on the bounding rectangle of the original nodes—only a few need to be considered for the minimal spanning tree (in our problem, only 31 out of a possible 936). Without such a reduction, the problem would be intractably large. The reduction follows from **Theorems 3** and **5**.

- We need at most $n - 2$ phantom points and $n - 1$ edges to minimize the spanning tree. This observation follows from **Theorem 4**.

Proofs of Preliminary Results

We are given a set of basic points $P = \{(x_i, y_i)|i = 1, 2, \ldots, n\}$. We define the sets of x- and y-coordinates of these points separately as $X = \{x_i\}$ and $Y = \{y_i\}$. Our goal is to form a connected graph on these points with lowest possible cost (dependent on length and station cost), where we are allowed to introduce "phantom stations" as we desire. Let the set of such new points in the optimal configuration be $P' = \{(x_i', y_i')|i = 1, 2, \ldots, m\}$, and as above put $X' = \{x_i'\}$ and $Y' = \{y_i'\}$.

Theorem 1. *Every configuration of connections and extra stations will always have a path length of at least $d = x_{max} - x_{min} + y_{max} - y_{min}$, where $x_{max} = \max x_i$, $y_{max} = \max y_i$, and so on.*

Proof: Since all of the basic points must be connected (possibly via extra stations), any vertical line $x = k$ with $x_{min} \le k \le x_{max}$ must cut the path in one or more points (otherwise we would have split a connected set into two disjoint pieces, which is a contradiction). Hence, if we project the actual path onto the x-axis, its length is at least $x_{max} - x_{min}$. Similarly, the projection of the path onto the y-axis is at least $y_{max} - y_{min}$. Any projection of the path will be less than its actual length (regardless of the metric, as long as edges are continuous). Moreover, since the path is composed only of horizontal and vertical segments (using rectilinear distances), the vertical components contribute nothing to the x-axis projection and vice versa; so the total path length has at least the length indicated.

Theorem 2. *There is always an optimal configuration.*

Proof: Given any set of n basic points, construct a spanning tree S by connecting P_1 to P_2 to $P_3 \ldots$ to P_n; this tree has some finite length d. With the given definition of cost, this configuration will have cost less than $d+nC$, where C is the cost of a degree-2 station. Using the rectilinear metric, there are finitely many configurations with length less than $d + nC$, and each of these will have a cost of at least its length (depending on station cost). Therefore, there is only a finite number of paths whose total cost is less than the cost of S. Hence, there must be a minimal-cost configuration.

Theorem 3. *For any optimal configuration (regardless of the cost of any extra stations), $X' \subset X$ and $Y' \subset Y$. (This says that all phantom stations must lie only on the same horizontal and vertical lines as the basic points.)*

Proof: (by induction) We need only show the first of the two results; the second follows identically. Let $X^* = X \cup X' \backslash \{x'_m\}$. In the optimal configuration, there is a certain set of $k \leq m + n$ points connected to the point (x'_m, y'_m). Let $\hat{X}^* = \{\hat{x}_i | i = 1, 2, \ldots, k\}$, where $\hat{x}_1 \leq \hat{x}_2 \leq \cdots \leq \hat{x}_k$ are the x-coordinates of these points. The cost of the overall configuration can be viewed as being made up of a contribution that depends on the position of phantom point P'_m and another that doesn't depend on this position (connections not involving P'_m and station costs). Thus, we can write the cost as $f(x'_m, y'_m) + C$, where C is a constant. In particular, since we are using rectilinear distances, we have:

$$f = f_{\hat{X}^*} + f_{\hat{Y}^*} = \sum_{\hat{x} \in \hat{X}^*} |\hat{x} - x'_m| + \sum_{\hat{y} \in \hat{Y}^*} |\hat{y} - y'_m|.$$

Consider only the first term. If $x'_m \notin \hat{X}^*$, then there exists $l \leq k$ such that

$$\hat{x}_1 \leq \hat{x}_2 \leq \cdots \leq \hat{x}_l < x'_m < \hat{x}_{l+1} \leq \cdots \leq \hat{x}_k.$$

Thus,

$$f_{\hat{X}^*} = \sum_{i=l+1}^{k} (\hat{x}_i - x'_m) + \sum_{i=1}^{l} (x'_m - \hat{x}_i).$$

In the case that $l < k/2$, we can reduce the value of this sum by setting $x'_m = \hat{x}_{l+1}$, as doing so will increase the value of each of the $l < k/2$ terms in the first sum by a fixed amount Δx but will reduce each of the $k - l > k/2$ terms in the second sum by the same amount. In the same way, if $l > k/2$, we could lower the sum by setting $x'_m = \hat{x}_l$ (note that if $l = k/2$, we can set $x'_m = \hat{x}_l$ and leave the value unchanged). However, this is a contradiction to the optimality of the configuration (changing the x-coordinate of one point doesn't affect the cost in any other way). Hence, we can assume that $x'_m \in \hat{X}^* \subset X^*$ as required. Proceeding by induction on all of the "movable" points (x'_i, y'_i), we must have $X' \subset X$ and similarly $Y' \subset Y$.

Corollary. *For n basic points, there are at most $n^2 - n$ locations for extra stations in an optimal configuration, given by $\{(x_i, y_j) | x_i \in X, y_j \in Y, 1 \le i, j \le n\}$. (There may be even fewer possibilities if some of the basic points lie on the same horizontal or vertical line.)*

Comment: We did not use the assumption that extra stations are at lattice points. Since all the basic points are lattice points, any extra points must be too.

Lemma 1. *For a network of n points (fixed or otherwise), the optimum spanning graph contains exactly $n - 1$ edges.*

Proof: We know that the optimum solution can have no closed cycles (if there were a loop, we could cut one of its edges, leaving a still connected graph but saving the cost of one whole edge). In other words, this graph is a tree. Therefore, there exists a point of degree one. We can remove this point and its connecting edge, leaving a connected set with one fewer edge and one fewer point, which again contains no closed loops. We continue in this way until we have one point left and no more edges. Hence, the tree contained only $n - 1$ edges.

Theorem 4. *At most $n - 2$ extra stations are required in an optimal configuration (the parallel of the given result for networks with a Euclidean metric).*

Proof: Suppose the optimal configuration (which exists by **Theorem 2**) has n basic points and m phantom stations. Each basic point must have degree at least one, since all points are connected. Further, each phantom station must have degree at least three or else there is no reason to have it in the configuration (a station of degree one doesn't connect anything, and a station of degree two just lies on a line). If we count all the edges emanating from each point in the graph, we will count all edges twice (since every edge has two ends). Thus, we arrive at the inequality $2e \ge 3m + n$, where e is the number of edges. But the number of edges in the optimal configuration is $e = m+n-1$. Thus, we get the desired result. (Alternatively, we could apply Euler's relation $V + F - E = 2$ with $F = 1$ and $V = m + n$.) Substituting in the inequality, we get $2(m + n - 1) \geq 3m + n$, or $m \leq n - 2$, as required.

Comment: Combining this result with the Corollary to **Theorem 3**, we see that we have at most $\binom{n^2-n}{n-2}$ possible combinations of the phantom stations. Since every configuration of basic points and extra points has a certain minimum spanning tree, we can actually use brute force to find the required configuration if we want. Note also that the given proof works for any metric—it doesn't use any specific measuring information, just connectivity.

Theorem 5. *For any point $P^* = (x^*, y^*)$ such that all basic points P_i satisfy $x_i \leq x^*$ or $y_i \leq y^*$, then there can be no phantom point P_i' with $x_i' \geq x^*$ and $y_i' \geq y^*$ in an optimal configuration. (Note that by symmetry we can get three other similar results by swapping either or both of the x or y inequalities.)*

Proof: We proceed by contradiction. Assume there is such a phantom station (x_i', y_i') in the optimal configuration. Clearly, no path terminates in the region $x \geq x^*$, $y \geq y^*$, since any such terminating path could be removed, leaving all of the basic points still connected at lower cost. Hence, the paths inside the region connected to (x_i', y_i') must meet the lines $x = x^*$, $y = y^*$ in the points $A_1(a_1, y^*), A_2(a_2, y^*), \ldots, A_q(a_q, y^*)$ and $B_1(x^*, b_1), \ldots, B_r(x^*, b_r)$, where $x^* \leq a_1 < a_2 < \cdots < a_q$ and $y^* \leq b_1 < \cdots < b_r$. Because no path terminates in the region, we must have $q + r \geq 2$. For the case of $q \geq 2$ and $r = 0$, it is clear as in **Theorem 1** that the length of the tree connected to (x_i', y_i') is at least $a_q - a_1$ by projection onto the line $y = y^*$. Hence, we could replace all of these connections with the segment (a_1, y^*) to (a_q, y^*) and have an equal or lower cost (since we need introduce no new stations, keep connections as they are, but get a shorter path). In the case $q = 0$ and $r \geq 2$, we use the segment (x^*, b_1) to (x^*, b_r); and for $q > 0$ and $r > 0$, we use the two segments (x^*, b_r) to (x^*, y^*) to (a_q, y^*). In all cases, the cost is equal or reduced.

Comment: This theorem allows us to reduce further the locations for phantom stations by chopping "corners" off the set of possible locations, as follows.

For each $y_k \in Y$ $(1 \leq k \leq n)$, set:

$$x_{00}(k) = \min_{\substack{(x_i, y_i) \in P \\ y_i < y_k}} \{x_i\} \qquad x_{01}(k) = \min_{\substack{(x_i, y_i) \in P \\ y_i > y_k}} \{x_i\}$$

$$x_{10}(k) = \max_{\substack{(x_i, y_i) \in P \\ y_i < y_k}} \{x_i\} \qquad x_{11}(k) = \max_{\substack{(x_i, y_i) \in P \\ y_i > y_k}} \{x_i\}.$$

Then by **Theorem 5** and the three symmetric results, there can be no phantom stations in any of the four regions:

$$x < x_{00}(k),\ y < y_k, \qquad\qquad x < x_{01}(k),\ y > y_k$$

$$x > x_{10}(k),\ y < y_k, \qquad\qquad x < x_{11}(k),\ y > y_k.$$

In our particular problem, we need only consider the 31 possible locations in **Figure 1** (p. 142), as opposed to the $9 \times 9 - 9 = 72$ points required using only the Corollary to **Theorem 3**.

Minimal Spanning Trees

Given a network of n points, we need to be able to construct the spanning tree of minimal total length. There are many algorithms to do this exactly, and we chose one that is fast and particularly adaptable for a computer. The minimum spanning tree algorithm of Minieka [1978] is guaranteed to find a tree (if one exists) in $\binom{n}{2}$ steps. Briefly, it works as follows:

The set of all paths connecting every point to every other point in the network and their lengths is created. The algorithm then selects from the set the next path with the shortest length (i.e., lowest cost) and will do either of two things:

- Reject the path if its inclusion will create a complete cycle with the rest of the paths already selected, or
- Include the path in the tree if it is not part of a complete cycle.

The tree is complete when it consists of $n - 1$ separate paths. Since there are exactly $\binom{n}{2}$ possible such paths in the network, the algorithm is guaranteed to find a solution in that many steps. Thus, we can now extract a minimum spanning tree for *any* set of points.

Note that this algorithm doesn't take into account station costs.

Strategy

Lacking an algorithm for finding a minimal rectilinear Steiner tree, we can use either a slow exact algorithm or a faster nonexact one:

- **Brute force:** We simply try all possibilities for both the number of phantom points and their locations. In our problem, using 9 fixed points with up to 7 phantom points at 31 possible locations would require

$$1 + \binom{31}{1} + \binom{31}{2} + \cdots + \binom{31}{7} = 3,572,224$$

tries. With each try taking approximately 0.017 sec (to calculate the minimum spanning tree and its length) on a 25-MHz 386-based PC, this is approximately a 17-hour operation. This method will give an absolute minimum configuration and can be adapted to any finite starting network configuration. Despite being highly inefficient, it will do the job for the given dataset. We ran our program simultaneously on six workstations (MIPS, Irises, Suns, and 386s), each doing different tries (a poor mathematician's imitation of parallel processing), to confirm the results we got with the second, more elegant approach. [EDITOR'S NOTE: We omit the program.]

- **Simulated annealing:** This is a more elegant way of searching out all the feasible combinations of phantom points and locations. We start with a given tree of phantom points and locations and allow the annealing program to create a new configuration. For each new configuration, the routine determines the minimum spanning tree and calculates the tree length. The annealing program then decides whether to retain or reject the new configuration based on a given *cooling schedule*. Modifying a given configuration and evaluating the modification are both very easy. The approach is completely general: Any starting configuration may be used, and the algorithm does not directly depend on the set of fixed network points. Simulated annealing, as opposed to the brute-force method, may not produce the absolute minimum. The chance that it will settle on a local minimum different from the absolute minimum can, however, be made as small as we wish, while still providing a very considerable time improvement over the first approach.

Part 1: No Station Costs

Figure 1 (p. 142) is a graph of the starting network of basic (or fixed) points, along with the reduced set P' of possible phantom-point locations. The program for simulated annealing (written in C) takes as input fixed points and the possible phantom-point locations, along with a starting path. [EDITOR'S NOTE: We omit the program.] (We started the program with an optimal path that we found by hand: We divided the fixed network into groups of three points, optimized each group by adding one phantom point, and then optimized the groups. The resulting configuration turned up several times as the minimal path of the annealing routine.)

The current configuration changes in a way randomly chosen from:

- adding a new phantom point randomly to one of the allowed locations,

- removing an existing phantom point, and

- moving an existing phantom point to a new random location.

(Note that in these three ways, we can move between all possible network configurations. For doing so, the third way is actually unnecessary; but we included it to give the annealing procedure more freedom.)

With the new configuration in memory, a calculation routine is called to set up the minimal spanning tree and evaluate the cost of the new network. This cost is then used by the annealing part of the program (routine METROP from Press et al. [1988, 351]) to decide whether the new configuration should be retained or rejected, according to an agreed-upon cooling schedule.

Notes of Interest: The first few runs indicated a trend to add phantom points of degree two, which are useless. Even though in Part 1 of the problem stations don't cost anything, we decided to eliminate this excess by the program by introducing a small penalty for each phantom point. The penalty was made small enough not to discourage adding phantom points where necessary but to eventually dispose of unnecessary ones.

We also added a feature not normally present in annealing routines, whereby the program remembers the best solution encountered during the entire run (simulated annealing usually only returns the last configuration used).

Results

- **Brute Force**

    - Brute force ran about 3.5 million calculations, occupying two hours of searching on each of six workstations.
    - Five distinct optimal networks were found, all of length 94; three had five phantom stations and two had four (**Figure 2**) (p. 143).

- **Simulated Annealing**

    - It took an average of about 9,800 iterations of the annealing routine (approx. 1.5 min using a 25-MHz 386-based PC) for convergence to an optimal solution.
    - Different runs, with different seeds to the random number generator, gave all of the five different optimum solutions of length 94.
    - In more than 100 runs, simulated annealing always converged to one of the five optimal paths, proving its worth for more general instances of the problem (when brute force is impossible).

Part 2: Station Costs

The problem is the same as Part 1, except for a cost factor for each station of $w(\text{degree})^{3/2}$, with $w = 1.2$. Two competing assumptions are:

- **Only phantom stations cost:** This is not realistic but can provide useful insight. Since each phantom station is of degree at least three, it will have a cost of at least $3^{3/2} \times 1.2 = 6.24$. Now, the length of the minimum path with phantom stations is 94 (from Part 1), but the length of the minimum path with phantom stations is only 110. Hence, to keep the cost under 110, one can afford at most two phantom stations. A quick brute-force calculation showed that no gains can be made by adding phantom stations, so the optimum tree is the minimum spanning tree with no phantom points.

- **All stations cost:** If we wanted to make use of the annealing routine developed for Part 1, we could either:
    - use some type of brute-force approach;
    - anneal the location, number, and degree of phantom stations (nested annealing!); or
    - modify the minimum spanning tree algorithm to try find the least-expensive (as opposed to shortest) tree, and use annealing as in Part 1.

It seemed that the first two options would be too time-consuming and inefficient, so we elected to try the third.

Modifying the Minimum-Spanning-Tree Algorithm

Given a set of points, we wish to construct a spanning tree of least cost, based on the cost formula given above. We base it on the routine described earlier, with the following changes:

1. All edges are calculated and their cost evaluated as $(\text{length} + 2w)$ (i.e., we assume for now a station of degree one at each end).

2. The least-cost edge that does not form a cycle joining points (A, B) is selected and put in the tree.

3. The cost of all edges containing either A or B is recalculated by incrementing the degree of the endpoint, since a station of degree one higher would now be created if we used such an edge. We then go back to Step 2.

As before, the routine ends when $(n - 1)$ edges have been selected, in no more than $\binom{n}{2}$ steps. With the modification, we can use both annealing and brute force exactly as before, with minimal code changes everywhere except in the actual spanning-tree generator.

The modified routine is only heuristic; it may not generate the least-cost spanning tree. However, we expect it to give a good spanning tree, an expectation confirmed by examining some of the trees it generates.

Modification Shortcomings

The new minimum-cost heuristic has some definite shortcomings:

- It does not guarantee the overall minimum cost path.

- Moreover, since the general problem is NP-complete [Chung and Whang 1979], we cannot possibly always solve it exactly in $\binom{n}{2}$ steps, as our procedure would seem to suggest.

- The routine is considerably slowed (by as much as a factor of 2) by having to recalculate costs at every step.

What this procedure does seem to do is to provide a very good (usually dead-on) approximation to the minimum-cost spanning tree, and the short-comings may be overcome by several restarts of the problem (though, as we shall see, this will not be necessary for the given problem).

Results

From several different annealing runs, we came up with two different trees with two phantom stations and cost 135.89 (see **Figure 3** on p. 144). (Note: The resulting network may sometime contain two edges that occupy the same lattice line. If such a result is not acceptable, one can always manually adjust the tree—without changing its total length or cost—to avoid it.)

We get a lower bound of 94 for the cost, from the minimum spanning tree with no station costs. Of the nine fixed points, at most eight can be of degree one (this is clearly underestimating the degree), for a total cost of 8×1.2, plus one additional one of degree two, costing $2^{1.5} \times 1.2 = 3.39$. Therefore, any spanning tree must cost at least $94 + 8(1.2) + 3.4 = 107$. Now, for every additional phantom point, we add at least the cost of a station of degree three, 6.24; hence, a network with m phantom points must cost at least $107 + 6.24m$. With our minimum from simulated annealing (clearly an upper bound for the absolute minimum) calculated at 135.89, we see that we can have at most four phantom stations with this cost scheme.

We stoop once more to the level of brute force and, using up to four phantom points, search out the minimum configuration (this search is much quicker than the previous brute-force calculation with up to seven). As before, the brute-force sledge-hammer verified our annealing results. Both the annealing and the brute-force results depend on the correctness of the modified minimum-spanning-tree routine but should in any case be very close to the truth.

Conclusions and Generalization

If we regard Part 1 as the special case $w = 0$ of Part 2, we then have two general procedures for optimizing a given spanning communication network. The simulated-annealing scheme produced verifiable results quickly, and it can accommodate any fixed-point network and any cost parameters. For a small number of points (or very fast vectorized/parallel computing machines), a brute-force approach is a not very elegant but absolute method of solution.

For $w \neq 0$, we can bound the number of phantom stations by using the optimal or near-optimal tree found for the case $w = 0$, and thus find upper and lower bounds for the minimum cost of any configuration using n or fewer phantom stations.

For the problem at hand, we found optimum solutions. We know these solutions to be absolutely optimal (for Part 1) or very likely to be optimal (Part 2), from brute-force checking.

While simulated annealing cannot does not always produce the absolute optimum solution, it did give the absolute minimum in every one of the more than 100 trial runs we did, indicating an acceptably high success rate.

Other Generalizations

If, instead of generalizing our point set or the value of w, we wish to make larger generalizations, then we can take one of several routes. First, we can generalize the station-cost function. Our near-optimal tree generation routine can be applied with just the cost function being changed. The brute-force technique, with the bound on the number of phantom stations, can also be applied.

On the other hand, we could generalize to networks located in more general spaces than $\mathbb{R}^2$ with the taxicab metric. Our techniques will work only for spaces in which we can find a bounding box with finitely many feasible sites for the phantom stations. This includes all of the spaces S^n with the taxicab metric, for $S \subset \mathbb{R}$, since **Theorems 3** and **5** can be generalized. We retain the results restricting feasible points to having coordinates from the projections of the set of stations, as well as the restriction to a bounding box with corners removed (as in the comment following **Theorem 5**).

We may generalize to any metric derived from $\| \cdot \|_p$, for $1 \leq p < \infty$; but then we require S to be discrete. In this case, not having the taxicab metric, we will have to check many more points in our bounding box, since we can no longer confine our attention to points whose coordinates lie in projections of the initial stations. In particular, we can easily generalize to $\mathbb{R}^n$ in the taxicab metric, as well as to $\mathbb{R}^n$ in any metric derived from a norm $\| \cdot \|_p$.

However, even with no greater number of basic points, still we will have many more feasible points than before. In the taxicab metric, with m initial stations and a space in ^{n}, we have now $\mathcal{O}(m^n)$ feasible locations for phantom stations; while for ^{n} in other metrics, the number of feasible locations is proportional to the hypervolume of the bounding box of the points. Thus, in some cases, the brute-force approach will be much more impractical. While simulated annealing can be expected to take longer to run, it should nonetheless work, although—since the state-space is much larger—it may no longer be as likely to give the optimal solution.

We could also generalize by adding extra cost criteria, such as minimal radius or diameter of the (weighted) graph representing the network (which may be useful if we wish to make the network respond faster), or having some preferential network routings. Unfortunately, such changes would render ineffective our heuristic approach to finding near-optimal spanning graphs for a fixed set of phantom stations and a fixed node-cost function based on degree. We would need either to develop a new heuristic approach for these specific cost criteria or to use a general method (such as simulated annealing) to find the near-optimal spanning graph for fixed phantom stations. As before, the phantom stations can be found by an outer annealing procedure or brute force, although if the spanning tree computations take too long, brute force would be completely impossible. Thus, we would have to resort to either a heuristic or an annealing procedure to find an optimal tree with fixed stations, and then an outer annealing to find the best configuration.

Impermeable or costly-to-cross obstacles, around which we must route the network, can be handled by modifying our metric (i.e., the metric function would route around impermeable obstacles and add the crossing cost to the distance; if done properly, doing so will still give a metric function). We would need to derive a bounding box for all the feasible network configurations; but this can be done fairly easily, since if we can find any spanning graph, a box of radius equal to the cost of this graph can be constructed. As long as we work in a discrete space such as ^{n}, we can use all of our machinery, though brute-force might be inefficient.

References

Chung, F.R.K., and F.K. Whang. 1979. The largest minimal rectilinear Steiner trees for a set of n points enclosed in a rectangle with given perimeter. *Networks* 9 (1979): 19–36.

Minieka, Edward. 1978. *Optimization Algorithms for Networks and Graphs*. New York: Dekker.

Press, William H., et al. 1988. *Numerical Recipes in C: The Art of Scientific Computing*. New York: Cambridge.

Judge's Commentary: The Outstanding Steiner Tree Papers

Jonathan P. Caulkins
Heinz School of Public Policy and Management
Carnegie Mellon University
Pittsburgh, PA 15213

The rectilinear minimum spanning tree problem in this year's competition is both intrinsically interesting and of practical importance. As such, it evinced a spectrum of responses which revealed the creative potential of inquisitive undergraduate students. Judging so many fine papers, therefore, was a challenging task. Broadly speaking, however, four factors distinguished the exceptional papers from those that were merely very good: maturity, respect for the problem, generality, and flexibility.

Maturity manifested itself in straightforward ways, such as a review of the relevant literature and a presentation of professional quality in both form and writing as well as mathematical substance. More importantly, the maturity of the superior teams was apparent in their approach to the problem. They fully understood the problem (evidenced, for example, by recognizing the differences between the Euclidean and rectilinear minimum spanning tree problems); they explicitly listed the assumptions underlying their analysis; they formally proved intermediate results; they communicated their intuition for the problem and their proposed solution(s); and they forthrightly described the limitations of their approach.

The best papers were also characterized by a healthy respect for the problem. Other papers spoke of "efficient algorithms for finding the optimal solutions." The better ones recognized that the rectilinear minimum spanning tree problem is NP-complete and were correspondingly more circumspect in their claims. When optimality is elusive, bounding arguments are a natural recourse. The relatively few teams that used bounding arguments distinguished themselves by doing so.

The utility of solutions often is decided by their generality. Some teams recognized and addressed the difficulty of generalizing their procedures to larger problems. Others overlooked the issue; still others offered only manual procedures that were not formally specified or relied extensively on human judgment or image-processing skills and hence could not easily be implemented on a computer for larger applications. Similarly, some teams embarked upon brute-force exhaustive searches. The clever ones pruned their search beforehand.

Flexibility often characterizes genius, and such was the case with this contest. Many teams found one procedure for obtaining a rectilinear spanning tree, documented it, and submitted their solution. More-creative teams developed two or more approaches and compared their merits. The best teams' comparisons included a sensitivity analysis with respect to both problem parameters and problem size.

It is worth noting that the four factors that differentiated various teams' performances do not represent hoops through which they must leap to win a prize; there is no such "formula for success" or "checklist of desiderata." Rather, they are signatures of quality that become most visible after the judging when one looks back over the field of competitors. Such characteristics help teams win contests, and this contest wins inasmuch as it helps stimulate such characteristics.

About the Author

Jonathan P. Caulkins did his undergraduate work at Washington University's School of Engineering and Applied Science, where he participated in the MCM. His teams' papers were rated Outstanding in the first two competitions, in 1985 and 1986.

Jon went on to earn a masters in electrical engineering and computer science and a doctorate in operations research from MIT. Now he is an assistant professor of operations research and public policy at Carnegie Mellon University's Heinz School of Public Policy and Management, where his research focuses on developing mathematical models of illicit drug markets.

Jon was an associate judge for the Steiner Tree Problem.

Practitioner's Commentary: The Outstanding Steiner Tree Papers

Marshall Bern
Xerox Palo Alto Research Center
3333 Coyote Hill Road
Palo Alto, CA 94304

Introduction

I'm more of a theoretician than a practitioner, but I have worked on Steiner tree algorithms and talked to a number of practitioners designing communications networks and computer chips. The rectilinear Steiner tree problem arises in VLSI design. An electrical network connecting a set of points on a chip should be as short as possible (more precisely, should have minimum capacitance) in order to reduce its charging and discharging time and increase its speed of operation.

A closely related version of the Steiner tree problem requires a network connecting a given subset of the stations within a much larger communications network. This version of the problem arises in pricing hookups for telephone customers with offices in a number of different cities. Thus the MCM Steiner tree problem posed a fairly realistic challenge.

Exact Algorithms

We might first attempt to solve the problem by brute force. We initially assume that all phantom stations, or *Steiner points*, lie on integer lattice points in the rectangle $[0, 35] \times [0, 24]$. There are 891 (that is, $36 \times 25 - 9$) possible Steiner-point locations in the rectangle. We cannot, however, test locations one at a time, because the best tree may not include the single best Steiner point. To be sure to include an optimal solution among our possibilities, we must test each subset of locations by computing the minimum spanning tree for that subset along with all the given stations. There are over 2×10^{20} subsets of locations with at most seven points. This is an appallingly large number, far too big to search.

A theorem of Hanan [1966], cited by all three Outstanding teams, states that we need only consider Steiner points that share their x-coordinate with one given station and their y-coordinate with another given station. In other words, we need only consider a grid of points formed by passing vertical and horizontal lines through the given stations. This reduces the number of possible locations to 63, and the number of subsets to $\sum_{k=1}^{7} \binom{63}{k}$, which is about 630,000,000.

A further reduction results from removing points near the corners of the grid. We can safely remove a corner point (that is, one at which the outer angle measures 270°) if it is not a given station; this rule may be repeated to produce a reduced grid with 31 possible locations and about 3,600,000 subsets. This reduction was used first by Yang and Wing [1972] and has been rediscovered several times, most recently by the team from the University of Western Ontario. We are now down to a manageable search, several hours on a fast workstation. The Western Ontario team used this approach (as well as a heuristic approach), thereby proving that the best solution to Part 1 of the problem has cost 94.

The brute-force method is not the only exact algorithm for the Steiner tree problem; there is a somewhat faster algorithm that uses dynamic programming [Dreyfus and Wagner 1972; Levin 1971]. This algorithm avoids testing all possible subsets of Steiner points, although it does have to compute optimal Steiner trees for all subsets of given stations.

Using some variation of this algorithm, it should be possible to find the exact solution to rectilinear Steiner tree problems with about 50 points in a few hours; but I do not know of anyone actually doing this. Computer-chip designers usually are content with approximate solutions, because the total length of a network is not always the limiting factor in the speed of operation. Furthermore, the Steiner tree problems arising in chip design are not very pure—other considerations, such as wiring around obstacles or leaving room for later wirings, may be important.

Approximate Algorithms

Because the computer time to solve the rectilinear Steiner tree problem exactly grows explosively, practitioners really use algorithms that find approximate solutions. One approximate algorithm is to use the minimum spanning tree, that is, to include no phantom stations at all. Hwang [1976] proved that this tree is never more than 50% longer than the optimal Steiner tree. This might be good enough in some practical situations, but it would not suffice as a contest solution.

All three teams programmed approximate algorithms. The team from Beloit College looked for the most advantageous single Steiner point and then added that point to the set of given stations. They repeated this "greedy" process until either the number of added stations reached seven or no further improvement was possible. This method found an optimal solution, though it will not always do so. (In their text, this team implies incorrectly that their algorithm is an exact algorithm.)

The team from Mount St. Mary's College programmed a number of different heuristics and admirably tested their methods on four problem instances besides the given one. They made another good decision in evaluating their algorithms; they report the relative improvement over the minimum spanning tree, as they cannot in general compute the exact Steiner tree. All of their heuristics made greedy decisions, that is, at each step they chose the cheapest alternative. They devised a "modified Kruskal" method that imitates a well-known algorithm for the minimum spanning tree problem but connects subtrees using Steiner points when advantageous. This heuristic found an optimal solution.

Though the Mount St. Mary team's heuristics have a speed advantage over the heuristic of the Beloit team, they probably do not produce solutions quite as good for larger problems. I am impressed that both of these teams invented quite reasonable approximate algorithms, very similar to those actually used in practice.

The Western Ontario team programmed a very different heuristic called "simulated annealing." Simulated annealing is a general scheme that randomly moves from solution to solution according to local rearrangement rules. The chance of moving to another solution depends on the costs of the two solutions and on a control variable called "temperature." With an appropriate "cooling schedule," the heuristic ultimately sticks at a nearly optimal solution. The team's simulated annealing program found an optimal solution 100 times out of 100 trials. I find this result surprising, and I wonder how well simulated annealing would do on larger problems.

Generalizations

Part 2 of the problem adds a new twist: each station has a cost that depends on its *degree*, that is, the number of lines meeting at the station. This assumption models situations such as telephone networks, in which stations as well as lines are expensive.

All three teams used heuristic algorithms for this problem, finding solutions of cost about 134.24 (Beloit), 134.85 (Mount St. Mary's), and 135.89 (Western Ontario). The Western Ontario team also performed an exhaustive search over all solutions with at most four Steiner points. There must have been a bug in their program, however, as the other teams beat them, and careful inspection reveals that their first solution (their **Figure 3**, p. 144) can be improved to 134.89 with a simple rearrangement. I do not know the optimal cost for Part 2, but I am sure all three teams came quite close.

Part 3 asked for further generalizations. Altogether, the Outstanding teams mentioned obstacles, alternative distance metrics, higher dimensions, and additional cost criteria. All of these are natural and useful generalizations [Hwang et al. 1992].

I shall elaborate a little on two of the more famous variants of the Steiner tree problem. The *Euclidean Steiner tree problem* uses the ordinary Euclidean metric. It is not hard to prove that, in an optimal solution, all Steiner points are incident to exactly three lines, and that these lines subtend angles of 120°. Even armed with this theorem, however, it is a hard problem to find a finite set of points which includes all possible locations of Steiner points. Z.A. Melzak [1961] gave a geometric construction that accomplishes this task; his paper really initiated modern research on the Steiner tree problem.

Another variant of the Steiner tree problem is the *phylogenetic Steiner problem*, proposed by Cavalli-Sforza and Edwards [1967]. In this variant, the given points represent organisms, and the optimal solution represents the most likely evolutionary tree relating these organisms. The minimum-cost criterion corresponds to a "parsimony principle": the most likely tree is the one requiring the fewest mutations.

Recent Research on Steiner Trees

Ronald Graham and I wrote an article for *Scientific American* on the Steiner tree problem less than three years ago [Bern and Graham 1989]. Already that article could stand an update.

The article's teaser asks, "What is the shortest network of line segments interconnecting an arbitrary set of, say, 100 points? The solution to this problem has eluded the fastest computers and sharpest mathematical minds." Later in 1989, Ernie Cockayne and Denton Hewgill [1992] of the University of Victoria sent us a picture of an optimal solution to a Euclidean Steiner tree problem on 100 random points. More-regular arrangements of points, such as points in a grid, tend to be harder, and a 100-station grid is probably (careful here!) still beyond the state of the art.

Later in the article, we mention the long-standing open problem of proving that for any set of points, a minimum spanning tree is never longer than $2/\sqrt{3}$ times the length of an optimal Euclidean Steiner tree. This conjecture was finally proved (very elegantly!) in 1990 by D.Z. Du and F.K. Hwang [1992].

The most recent big theoretical results on Steiner trees are contained in a sequence of papers starting with Zelikovsky [1993]. These papers give fast approximate algorithms with performance guarantees better than the guarantees offered by the minimum spanning tree. For example, P. Berman and V. Ramaiye [1994] have shown that although a minimum spanning tree may be 3/2 times as long as a optimal rectilinear Steiner tree, a certain greedy solution will never be more than 97/72 times as long.

For a survey of the latest results on the Steiner tree problem and many other hard geometric problems, see Bern and Eppstein [1995].

References

Berman, P., and V. Ramaiyer. 1994. Improved approximation algorithms for the Steiner tree problem. *Journal of Algorithms* 17: 381–408.

Bern, M.W., and R.L. Graham. 1989. The shortest-network problem. *Scientific American* 260(1) (January 1989): 84–89.

Bern, M.W., and D. Eppstein. 1995. Approximation algorithms for geometric problems. To appear in *Approximation Algorithms for NP-Complete Problems*, edited by D. Hochbaum. PWS Publications.

Cavalli-Sforza, L.L., and A.W.F. Edwards. 1967. Phylogenetic analysis: models and estimation procedures. *American Journal of Human Genetics* 19: 233–257.

Cockayne, E.J., and D.E. Hewgill. 1992. Improved computation of plane Steiner minimal trees. *Algorithmica* 7: 219–229.

Dreyfus, S.E., and R.A. Wagner. 1972. The Steiner problem in graphs. *Networks* 1: 195–208.

Du, D.Z., and F.K. Hwang. 1992. A proof of Gilbert-Pollak's conjecture on the Steiner ratio. *Algorithmica* 7: 121–135.

Hanan, M. 1966. On Steiner's problem with rectilinear distance. *SIAM Journal of Applied Mathematics* 14: 255–265.

Hwang, F.K. 1976. On Steiner minimal trees with rectilinear distance. *SIAM Journal of Applied Mathematics* 30: 104–114.

Hwang, F.K., D.S. Richards, and P. Winter. 1992. *The Steiner Tree Problem*. New York: North-Holland.

Levin, A.J. 1971. Algorithm for the shortest connection of a group of graph vertices. *Soviet Mathematics Doklady* 12: 1477–1481.

Melzak, Z.A. 1961. On the problem of Steiner. *Canadian Mathematical Bulletin* 4: 143–148.

Yang, Y.Y., and O. Wing. 1972. Optimal and suboptimal solution algorithms for the wiring problem. *Proceedings of the IEEE Symposium on Integrated Circuit Theory* (1972): 154–158.

Zelikovsky, A.Z. 1993. An 11/6-approximation algorithm for the network Steiner problem. *Algorithmica* 9 (5): 463–470.

About the Author

Marshall Bern is a member of the research staff at Xerox Palo Alto Research Center. He received a master's degree in mathematics from the University of Texas at Austin in 1980 and a Ph.D. in computer science from the

University of California at Berkeley in 1987. In between graduate schools, he worked in the area of signal processing. His current research is on algorithms for computer graphics and finite-element mesh generation—in fact, on any problems for which he can scribble pictures.

1992: The Emergency Power-Restoration Problem

Power companies serving coastal regions must have emergency-response systems for power outages due to storms. Such systems require the input of data that allow the time and cost required for restoration to be estimated and the "value" of the outage judged by objective criteria. In the past, Hypothetical Electric Company (HECO) has been criticized in the media for its lack of a prioritization scheme.

You are a consultant to HECO power company. HECO possesses a computerized database with real-time access to service calls that currently require the information:

- time of report,

- type of requestor,

- estimated number of people affected, and

- location (x, y).

Crew sites are located at coordinates $(0, 0)$ and $(40, 40)$, where x and y are in miles. The region serviced by HECO is within $-65 < x < 65$ and $-50 < y < 50$. The region is largely metropolitan with an excellent road network. Crews must return to their dispatch site only at the beginning and end of shift. Company policy requires that no work be initiated until the storm leaves the area, unless the facility is a commuter railroad or hospital, which may be processed immediately if crews are available.

HECO has hired you to develop the objective criteria and schedule the work for the storm restoration requirements listed in **Table 1** using the work force described in **Table 2**. Note that the first call was received at 4:20 A.M. and that the storm left the area at 6:00 A.M. Also note that many outages were not reported until much later in the day.

HECO has asked for a technical report for their purposes and an "executive summary" in laymen's terms that can be presented to the media. Further, they would like recommendations for the future. To determine your prioritized scheduling system, you will have to make additional assumptions. Detail those assumptions. In the future, you may desire additional data. If so, detail the information desired.

Table 1.
Storm restoration requirements.

Time (A.M.)	Location	Type	# Affected	Estimated Repair Time (hrs for crew)
4:20	(−10, 30)	Business (cable TV)	?	6
5:30	(3, 3)	Residential	20	7
5:35	(20, 5)	Business (hospital)	240	8
5:55	(−10, 5)	Business (railroad sys.)	25 workers; 75,000 commuters	5
6:00	All-clear given; storm leaves area; crews can be dispatched			
6:05	(13, 30)	Residential	45	2
6:06	(5, 20)	Area*	2000	7
6:08	(60, 45)	Residential	?	9
6:09	(1, 10)	Government (city hall)	?	7
6:15	(5, 20)	Business (shopping mall)	200 workers	5
6:20	(5, −25)	Government (fire dept.)	15 workers	3
6:20	(12, 18)	Residential	350	6
6:22	(7, 10)	Area*	400	12
6:25	(−1, 19)	Industry (newspaper co.)	190	10
6:40	(−20, −19)	Industry (factory)	395	7
6:55	(−1, 30)	Area*	?	6
7:00	(−20, 30)	Government (high school)	1200 students	3
7:00	(40, 20)	Government (elementary school)	1700	?
7:00	(7, −20)	Business (restaurant)	25	12
7:00	(8, −23)	Government (police station & jail)	125	7
7:05	(25, 15)	Government (elementary school)	1900	5
7:10	(−10, −10)	Residential	?	9
7:10	(−1, 2)	Government (college)	3000	8
7:10	(8, −25)	Industry (computer manuf.)	450 workers	5
7:10	(18, 55)	Residential	350	10
7:20	(7, 35)	Area*	400	9
7:45	(20, 0)	Residential	800	5
7:50	(−6, 30)	Business (hospital)	300	5
8:15	(0, 40)	Business (several stores)	50	6
8:20	(15, −25)	Government (traffic lights)	?	3
8:35	(−20, −35)	Business (bank)	20	5
8:50	(47, 30)	Residential	40	?
9:50	(55, 50)	Residential	?	12
10:30	(−18, −35)	Residential	10	10
10:30	(−1, 50)	Business (civic center)	150	5
10:35	(−7, −8)	Business (airport)	350 workers	4
10:50	(5, −25)	Government (fire dept.)	15	5
11:30	(8, 20)	Area*	300	12

*Area signifies a combination of two or more of the other classification types.

Table 2.

Crew descriptions.

- Dispatch locations at $(0, 0)$ and $(40, 40)$.

- Crews consist of three trained workers.

- Crews report to the dispatch location only at the beginning and end of their shifts.

- One crew is scheduled for duty at all times on jobs assigned to each dispatch location. These crews would normally be performing routine assignments. Until the "storm leaves the area," they can be dispatched for "emergencies" only.

- Crews work 8-hr shifts.

- There are six crew teams available at each location.

- Crews can work only one overtime shift in a work day and receive time-and-a-half for overtime.

Comments by the Contest Director

The problem was contributed by Joseph Malkevitch (Dept. of Mathematics and Computer Science, York College (CUNY), Jamaica, NY).

Development of an Emergency-Response System

Mike Oehrtman
Jennifer Williams
Kevin Yoder
Oklahoma State University
Stillwater, OK 74078

Advisor: James R. Choike

Introduction

The Hypothetical Electric Company (HECO) is a coastal power company that must deal with emergency power outages from time to time. In the past, HECO has been criticized for its lack of a prioritization scheme in handling emergency outage calls. As a result, our team has been hired to provide HECO with a technical plan for defining what objective criteria should be used in assessing emergency calls and using this information to schedule the required power-restoration work. While data from a particular storm (Hurricane Jane) were provided, we will provide a model broad enough to serve in many emergency situations.

We considered the following criteria:

- immediate response to emergency calls (hospitals and commuter railroads),

- minimization of overtime hours,

- minimization of travel time, and

- minimization of the number of people waiting for service at any time.

We developed a flexible model that uses the first, third, and fourth criteria and processed scenarios with varying overtime limitations. In our judgment, the best scenario for the given data and our assumptions uses each crew for 16 hrs/day. All emergency calls were finished in 8 hrs 10 min of the first call of this type. A total of 595 regular hours and 525 overtime hours were used, and 358 hours were spent driving. The final call was completed after 58 hrs, at a cost of $21,800.

Once all factors have been converted into an equivalent number of people affected, every job becomes quantitatively weighted. Combining this prioritization with localized considerations of distance and repair time results in an optimal dispatching algorithm. This model and the program designed to facilitate its application are adjustable to changing prioritization and to individual company objectives.

Clarification of the Problem

The first issue is the qualitative nature of the information provided by an emergency report. The problem requires a quantification of these data which will allow the dispatcher to weigh objectively, for example, a large number of people affected only a short while against a handful of workers who might wait for a long period of time. Additional data at the time of the report, as well as an understanding of city management, may help in this quantification.

The problem is formidable, because of the number of factors involved at every stage of the scheduling and decision-making process. Crew work times, including overtime considerations, must be weighed. Thus, an efficient utilization of all available crews is required. Distance to and between work sites is a large factor, since driving time consumes potential work time. However, the prioritization of outage sites clearly distinguishes this problem from a basic time-optimization problem. We concluded that an efficient, workable prioritization scheme is the primary goal, with cost vs. time being a secondary consideration to be determined by the company's desire and ability to sacrifice monetary savings for time of completion.

Assumptions Given

- The prioritization scheme required by HECO is for use during emergency situations only; HECO scheduling for routine days is not under consideration.

- By HECO policy, no work may be initiated until the storm leaves the area, with the exception of that required by a hospital or commuter railroad.

- The region serviced by HECO is metropolitan, and an excellent road network is available.

- Dispatch locations are at $(0,0)$ and $(40, 40)$ for the HECO service area within $-65 < x < 65$ and $-50 < y < 50$.

- Crews are required to report to dispatch sites only at the beginning and end of each shift.

- Only one crew is scheduled for duty at any time, with the implication that, during an emergency, crews that find themselves not immediately needed may go off duty.

- Each crew consists of three trained workers; six crews are available at each dispatch location.

- Crews work 8-hr shifts, can work only one overtime shift in a work day, and receive time-and-a-half for overtime.

Additional Assumptions

- Crews are available immediately upon request at the dispatch location.

- Crews must return to the same dispatch location from which they started and are uniquely assigned to that dispatch location.

- All distances are measured rectilinearly.

- Crews travel at an average speed of 30 mph between work sites.

- Crew trucks do not break down.

- A power outage maintains its original severity even when fractional parts of the work have been completed.

Assumptions Concerning Parameters

- All estimates given for unknown "# Affected" entries in **Table 1** of the problem statement are taken from averages of calls of a similar type. In the cases of "Government (city hall)" and "Government (traffic lights)," no calls of a similar type were available. Therefore, we specified 0 as the number affected.

- All estimates given for unknown "Estimated Repair Time" entries are taken from the average of all calls recorded from 4:30 A.M. to 11:30 A.M.

Background: Existing Models

The first stage in our analysis of HECO's emergency-response system was a survey of existing networking, minimizing, and prioritizing models. We did not find a precise existing model for such emergency-response networks; however, we outline the following four categories to clarify the pros and cons of applying elements of these models to the problem at hand.

Graph Theory

Graph theory provides a variety of models and optimization algorithms and is most heavily drawn upon in our model. The most relevant models are those beginning with a connected graph (all worksites are connected by roads to other worksites). Notice that the HECO model, as a continuously changing system, will require a graph that is updated continuously (or quasicontinuously, as there is not a decision to be made at every point in time). Distance, time, priority weightings, or a combination of these may be placed as weights upon the edges of the graph. Note that values that are inversely related, such as distance and time, will have to be adjusted to contribute in the same direction to the weight of an edge. Then a variety of minimal-spanning-tree algorithms may be applied to minimize the total time (waiting time, cost, etc.) required.

Prim's algorithm and Kruskal's algorithm [Jackson and Thoro 1990, 191–193], yield minimal spanning *trees*. The crews must return to the dispatch centers at the end of each shift, so a minimal spanning *cycle* would be more appropriate. Dijkstra's algorithm [Skvarcius and Robinson 1986, 231–233] utilizes a weight matrix and, since it provides a minimal path between two vertices, may be used to provide a minimal path from a point back to itself. However, this algorithm requires the consideration of every possible vertex in comparison to all others and would only be capable of calculating one path at a time. The key elements of graph theory minimization which are retained in our HECO model are:

- Weighting of paths on a graph provides a quick reference for minimization.

- Minimal path algorithms utilizing these weights are localized algorithms. In other words, from any vertex the next step in a minimal path may be found by minimizing that step.

The Critical Path Method (CPM)

The critical path method is an algorithm suited to large projects in which some activities may be pursued concurrently while others are dependent upon completion of particular predecessors. It is used primarily when a priority of completion is required and a deadline is set for completion of all activities. Notice that the prioritization required in the HECO response system may be thought of as a system of prerequisites, in which high-priority calls are required to be completed before low-priority calls. Thus, we may have a network as shown in **Figure 1**.

The CPM algorithm then uses this ordering and interdependence, along with time requisites for each task, and produces three variables of importance:

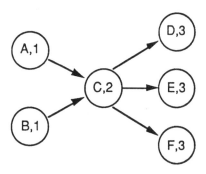

Figure 1. Network of jobs A, . . . , F, each of priority 1, 2, or 3.

- The earliest time of completion $e(x)$ for each job x. In the **Figure 1**, if A requires 3 hours and B requires 2 hours, then $e(C) = 3$, the time when all level-1 jobs are completed so that level-2 jobs may be started.

- The latest event time $l(x)$ for starting job x and still completing the entire project on schedule.

- Float time, $f(x) = l(x) - e(x)$, a range during which a job may be begun.

These three numbers are utilized by CPM to provide an efficient schedule [Minieka 1978, 319–330]. Notice that $l(x)$ requires an explicit deadline. Although it would be feasible to develop a deadline by which all repairs may be expected to be finished, we assume that HECO will wish to complete the entire project in minimal time. The following elements may be applied to the HECO model:

- Tasks of high priority may be considered prerequisites to those of lower priorities.

- A range of time (modeled after float time) may be useful in calculating whether a specific crew should or should not assume a job. Notice that crews themselves have deadlines, specifically the ends of their regular or overtime shifts.

Hypercube Queuing Model (HQM)

The Hypercube Queueing Model [Larson 1978], developed primarily to aid police dispatching, in many ways resembles the HECO model. HQM uses data on time of travel and service times. Mobile response units, whose locations are known at all times, are available to respond to calls. One unit is required per call. However, the model also requires that travel time be insignificant compared to service time, which is highly unrealistic for HECO, as HECO's service area comprises 13,000 square miles. Also, police patrol is essentially different, too, since patrols are required during no-call times. HQM breaks its service areas into small units and groups them into subunits based upon statistics of prediction for calls. Also, HQM cannot remove calls from a queue in any prioritizing manner nor interrupt any en-route or busy crew, and Larson [1978, 6] suggests that "If the system planner wishes to analyze these types of operation . . . , he should probably use a simulation model." Thus, for HECO we find the following.

- A useful means of tracking the status of units is with an array $(p_1, \ldots, p_n)$, where p_i encodes the status of unit i. A status of "in transit" may be useful.

- A simulation model may be useful in analyzing a queue in which items are to be removed and assigned based on a preemptive priority system.

Queuing Theory

Queueing theory provides many algorithms for analyzing a queue, based primarily on average rates as calculated, for example, by the Poisson formula for probability of an occurrence over continuous time. An algorithm of this type is (Mi/Gi/1):(NPRP/∞/∞), which calculates expected waiting times for a customer in a system and an individual queue, as well as the expected number of customers in the system or queue at a given time [Taha 1987, 633].

But note that these are algorithms that *predict* solutions that may be useful to HECO should it decide, for example, to relocate a dispatch center or to predict the number of crews that will be needed during a stormy season. What is required at the moment, however, is a more immediate and less statistical system for dispatching calls as they are received.

Analysis of the Problem

In analyzing the need for a prioritization scheme for use by HECO during emergency power restoration, we determined that the following goals are the most important:

- Immediate response to crucial outages (hospital and commuter railroads).

- Minimization of overtime hours, as a cost-saving strategy.

- Minimization of travel time between worksites, as a cost- and time-saving strategy.

- Minimization of the time that facilities must wait for repairs, counted from the time an outage is reported.

- Minimization of the number of people waiting for repairs at any time.

Not all of these goals may be directly optimized within the model, since they conflict (e.g., it is impossible to absolutely minimize both the number of overtime hours used and the waiting time for facilities).

What we desire, then, is a model that optimizes each quantity under the restriction of the others, i.e., uses these criteria to schedule objectively and repair each outage with a balance between cost and time.

Design of Model

General Groupings

The first step in the design of a model for scheduling crews to worksites is a grouping system of those sites. This grouping should primarily satisfy goal of immediate response to crucial outages. Within groups, we assign rankings. [EDITOR'S NOTE: We omit the table of rankings for space reasons.] Following is a general description of each group and its distinguishing characteristics.

- Group I: Facilities of utmost importance. Crews may be assigned to Group I facilities while the storm is in progress as well as after it leaves the area.

- Group II: Facilities needed for public safety. Crews may be assigned to Group II facilities only after the storm leaves the area.

- Group III: Facilities least crucial to general public safety. Crews may be assigned to Group III facilities only after the storm leaves the area.

Constants and Formulas

k (people per hour):
Description: k ensures that (all else held constant) call B (9:00) will be processed before call A (8:00) if and only if

(# of people affected at B) > (# of people affected at A) + k.

Rationale: Some constant must relate the relative worth of people affected to the time spent waiting, to provide a balance between the design goals outlined in the section on Analysis of the Problem. We set $k = 15$.

GW (Group Weight, converted to equivalent number of people affected):
Description: GW assures that (all else held constant)

- Worksites from different groups will be processed according to their group.
- Worksites within a group will be processed according to their relative importance. For example, in Group III, a residential area (GW = 0) would be processed before city hall (GW = 300) if and only if

(# of people affected in residential area) >

(# of people affected by city hall outage) + 300.

Rationale: Some constant must relate the relative importance of within-group worksites while at the same time maintain prioritization by group. Notice that GW allows us to pursue the goal of immediate response to crucial outages from the analysis of the problem.

PV

Description: PV (performance value) establishes a "value" in units of people for each call received by HECO. The considered factors are the time a call was received, the number of people affected, and constants k and GW. Note: The factor (24 − time of call) assumes that all calls come in during the first 24-hr period, and that time is measured in 24-hr time.

$$PV = (24 - \text{Time of Call}) \times k + (\text{\# of people affected}) + GW.$$

Rationale: Some cumulative criterion must be available to evaluate the importance of a call and determine its relation to other calls on the queue; PV will allow the model to balance the many desired considerations. Note: A Group I worksite will always have a greater PV than a Group II or III worksite, and a Group II worksite will always have a greater PV than a Group III worksite. This is achieved by setting GW values at appropriate ranges within each group.

SCORE

Description: SCORE is a value that is used when a crew is available to work and a decision must be made regarding where the next assignment should be. SCORE will change for a particular worksite depending upon the location of the crew in question. In simple terms, SCORE is equal to the site's PV times the percentage of the work that the crew would be able to complete.

$$SCORE = PV \times \left[\frac{\min \left\{ \begin{array}{l} \text{time required to finish job} \\ \text{time left on shift} - \text{time to travel to worksite and back to base} \end{array} \right.}{\text{time required to finish the job}} \right]$$

Rationale: Some criterion must be available that takes into account not only PV but amount able to be accomplished at a particular worksite. We chose the percentage of work able to be completed as the deciding factor.

Algorithm

[EDITOR'S NOTE: An extended description of the details of the algorithm is omitted for space reasons, as is the computer code itself.]

Determination of an Optimal Priority Scheme

Since all decisions regarding selection of worksites occur in the Crew Finish procedure, this is where we decided to concentrate on manipulation of the model. Five subroutines (based on slightly varying Priority Schemes) were constructed for the Crew Finish procedure, and the five resulting models were compared under the following constant conditions:

- $k = 15$,

- GW as assigned by us, and

- all crews worked two consecutive 8-hr shifts (one regular, one overtime) per work day for the duration of the emergency.

The five Priority Schemes are:

1. The first procedure models the manner in which HECO probably actually operated during Hurricane Jane. Crews are dispatched to sites as calls are received and are placed on a job list once no more crews are available. No consideration is given to criteria other than distance of possible worksites. When a crew becomes free and must decide between two or more jobs on the job list, it simply chooses the closest site.

2. The second procedure is a slight refinement of the first. A priority scheme (using PV) is applied to the list of jobs, ranking them in importance. When a call is received, it is assigned to a crew according to its PV and the distance of each crew from the worksite.

3. Once again, calls are assigned according to PV and crew distance. When a job must be selected for an idle crew, each job on the job list is considered. The percentage of the work which could be completed by the crew is computed, based on considerations of the amount of time left on the crew's shift, the travel time to the site, and back to the crew's base station, and the estimated size of the job. A SCORE value for each site is then calculated. The site with the highest SCORE value is selected.

4. The intent of the fourth procedure is to expedite service to locations on the job list with highest priority. When a crew becomes idle, SCORE values are calculated for each worksite on the job list, as in the third procedure, and the site with the largest value is deemed "first choice." "First choice" sites are then produced for each of the other 11 crews for the times those

crews would next need to make that decision. If any other crew would select the same job and arrive at the location faster, the initial crew must consider another choice (next highest SCORE value) and make another decision. If necessary, third, fourth, etc., choices are made. If no choice is acceptable, the crew returns to its base.

5. In many cases, when the fourth procedure was used, crews were being sent back to their base and left with nothing to do. In an attempt to utilize this time, the procedure was modified so that if all of a crew's possible choices turn out to be better suited to other crews, it returns to its "first choice."

Table 3 presents the data used for determination of the optimal model. The bases for the cost figures in the table are regular hours at $15/hr and overtime hours at $22.50/hr, driving at 30 mph, and truck costs of $0.30/mi. Note that with three members to a crew, the total worker hours spent driving must be divided by 3 to give the time of the truck on the road.

Table 3.
Comparing priority schemes.

Criteria	Priority Scheme				
	1	2	3	4	5
Group I sites done	19:07	14:25	14: 25	14:25	14:25
Group II sites done	41:19	17:07	17:07	17:07	17:07
Group III sites done	53:11	59:04	58:26	65:45	57:50
All sites done	53:11	59:04	58:26	65:45	57:50
Regular hrs	598	676	629	605	613
Total overtime hrs	483	550	498	474	512
Time working	762	762	762	762	762
Time driving	319	464	365	317	363
Cost	$20,800	$23,900	$21,700	$20,700	$21,800

From **Table 1**, we determined that Priority Scheme 5 best met the criteria that we had established at the beginning of the model-building process. For the remaining analysis, this will be the optimal model referred to.

Strengths and Weaknesses of the Model

Our model pursues all of the goals announced in the section on Analysis of the Problem. Every goal corresponds to a summand in the calculation of the Priority Value or in the calculation of SCORE.

There is an intrinsic weakness in the model linked directly to the nature of the problem. Because there is an element of subjectivity to any prioritization scheme, a "true optimal value" does not exist for even one of the variables. For example, brute-force simulation may produce a minimal total travel time by considering every possible combination of assignments; but this is the "optimal value" only if minimum travel time is the sole factor considered. However, directly from this weakness springs this model's primary strength. Recall that we concluded that a balance would be optimal. By varying the parameters in the algorithm, one can compare precisely those values that, in an individual case or to a particular company, represent an "optimal balance."

In addition, a functioning dispatch center will contribute many more variables that may complicate the model but will also facilitate some considerations. For example, many crews in our final model run left sites only 5 min before the job was estimated to have been completed. Most human beings would either finish the job or would have worked a little faster in order to meet a departure deadline.

Suggestions to HECO

We compared the applications of five different prioritization schemes to the established model. These results are based entirely upon tests run on the data provided by HECO concerning Hurricane Jane. Thus, our conclusion that Priority Scheme 5 is optimal would be strengthened by repeated application of the same test to data from different storms. By varying the input, the consistency of the model itself is tested and the prioritization schemes become more finely tuned.

HECO will wish to examine our tables in detail, noting specific company preferences for the results of varying Priority Schemes 1–5 and allowable employee overtime. In response to public dissatisfaction, we recommend that cost not weigh heavily in HECO decisions; however, cost information is provided, as it must certainly be a consideration for any profit-oriented organization.

The model requires an adjustment to enable the anticipation of a crew's departure from an unfinished job. Currently, there is a lag from the time of crew departure and the arrival of the new crew.

During Hurricane Jane, three calls were processed whose locations are outside the HECO service area. As purely additional data with which to fine-tune the developing model and prioritizations, we considered these sites. However, HECO must define its policy in perhaps one of the following ways:

- reject outside calls,

- respond within a predetermined extended perimeter, or

- respond dependent upon changing variables (e.g., proximity to outages within the perimeter, availability of non-overtime crews, etc.).

Any one of these policies may be incorporated into the algorithm and computer program.

Should HECO desire in the future to predict numbers of calls received, areas of heavy call traffic, waiting times on the queue, number of jobs likely to be in the queue, etc., it should pursue a more probabilistic model. These statistics are particularly useful for such tasks as relocating dispatch centers, hiring in anticipation of need, and expanding computer networks or dispatching systems. Queueing theory provides the most highly developed models, particularly when, as in the HECO problem, customers must be handled according to priorities.

Conclusion

Using the available data from HECO, our team defined objective criteria to prioritize emergency power-outage calls and produced a model to schedule the emergency work to available crews. Ideas taken from graph theory, queue theory, and hypercube queuing theory were used in our formulation and assumptions. The chosen optimal model is flexible enough that parameter changes (constants, formulas, amount of overtime allowed) can be easily implemented. HECO will be able to use this model to quantify and assess any number of scenarios, with a wide range of possible variables for future study. In short, service can improve and operation costs can decline if the information in this report is properly applied.

References

Jackson, Bradley W., and Dmitri Thoro. 1990. *Applied Combinatorics with Problem Solving*. Reading, MA: Addison-Wesley.

Larson, Richard C., ed. 1978. *Police Deployment*. Lexington, MA: Lexington Books.

Minieka, Edward. 1978. *Optimization Algorithms for Networks and Graphs*. New York: Marcel Dekker.

Skvarcius, Romualdas, and William B. Robinson. 1986. *Discrete Mathematics with Computer Science Applications*. Menlo Park, CA: Benjamin/Cummings.

Taha, Hamdy A. 1987. *Operations Research: An Introduction*. New York: Macmillan.

Judge's Commentary: The Outstanding Emergency Power-Restoration Papers

Jonathan P. Caulkins
Heinz School of Public Policy and Management
Carnegie Mellon University
Pittsburgh, PA 15213–3890

The Emergency Power-Restoration problem was truly challenging. The problem statement left many issues incompletely specified. These ambiguities, coupled with the very nature of the problem, allowed considerable latitude and hence demanded considerable modeling.

This open-ended character, which made the problem so interesting and so challenging for the participants, also made the problem interesting and challenging to judge. We clearly could not simply crown as winner the team whose schedule finished first, because there are multiple objectives and every team used its own relative weighting of these objectives. Furthermore, using different assumptions leads to different numerical results; it clearly would not be reasonable to compare directly the time to completion for a team that assumed crews traveled 60 mph with one that assumed crews traveled only 30 mph. Thus, there is no one "right" answer or single "optimal" solution.

Instead, the teams were evaluated based on how they approached the problem. We considered more factors than can possibly be detailed here, and every paper was judged individually; so it would be misleading to create the impression that winning teams simply needed to satisfy a checklist. Nevertheless, I will describe some characteristics that were associated with successful papers.

Perhaps the most important of these was *generality*. A power company would be unlikely to seek a schedule for one particular storm; by the time the consultants had the data for that storm, many (if not all) of the allocation decisions for that storm would have already been made. The power company is interested instead in objective criteria that it can use to schedule work crews during and after future storms and—as evidence of the viability of those criteria—wishes to see them applied to the given data. Teams that gave a schedule only for this one storm did not deliver as useful a product as did teams that gave a general procedure. Likewise, some teams implicitly or explicitly assumed that all storms would be similar to the one for which data are given; nature offers no such guarantees.

Since this was a contest in modeling, and models are simplified representations of reality, we also evaluated the reasonableness of the simplifying assumptions made and how well they were justified. Weaker papers simply asserted that a factor was unimportant, or worse, neglected even to identify the assumption. Stronger papers offered a rationale and discussed the implications of making or not making the assumption. Some papers distinguished themselves by performing sensitivity analyses and/or using bounding arguments to demonstrate quantitatively that the simplification made would not lead to substantially inferior solutions.

Another touchstone of quality was explicit recognition of the existence of multiple, often conflicting, objectives. Essentially every team recognized the choices to be made between different types of customers, but other factors that might enter into a priority calculation include the number of people affected and the elapsed time since the outage was reported. The "justice" of responding to calls in the order in which they are received must be balanced against the "efficiency" of dispatching crews to nearby locations (to minimize travel time) and the "wisdom" of giving priority to certain classes of customers. Since there are multiple objectives, and there is not one universally accepted way of converting the vector of "scores" on the corresponding attributes into a scalar measure of overall quality, it is expedient to report a variety of performance measures. The team from Oklahoma State University distinguished itself in this regard.

Another characteristic of a strong paper was the consideration of a variety of scheduling procedures. Many papers proposed a single approach, applied it, and reported the results in isolation. It is essentially impossible to ascertain whether such a procedure is good or bad. All the procedures could be labeled "bad," because at least a few customers were without power for a considerable time; likewise, all the procedures might be deemed "good," because they restored power to every customer eventually. Such statements are vacuous. It is only by comparing relative performance that one can hope to make useful statements about quality. Both the Washington University and the Oklahoma State University teams were outstanding in their use of five different approaches, including one (first come, first serve) that they inferred from the problem statement was likely to be similar to the power company's current practice.

Papers also varied with respect to the degree of sophistication of their scheduling algorithms. Many used a simple greedy heuristic that assigned the highest-priority not-yet-assigned job to a crew when the crew became available. Others were more clever. Some looked ahead before dispatching to see whether there were any crews closer to the destination who would become free shortly and hence be able to reach the scene first. Others used measures of shadow cost or opportunity cost to make assignments. One team even considered that working full 16-hr double-shifts can *extend* the completion time (the North Carolina School team's result not withstanding) if the crew cannot finish the final task that day, forcing additional travel time and delaying the earliest time at which the crew can be back on the job.

Likewise, some teams recognized that if the jobs are viewed as nonpreemptive, it may be prudent to leave a crew idle rather than immediately dispatch it, in case a higher priority call is received. The North Carolina School team considered an extreme version of this strategy, namely, holding back all crews for a certain amount of time. They performed a sensitivity analysis and calculated the optimal holding time for this storm; it is not zero. This team was also one of the relatively few that considered the possibility that the estimated repair times were only estimates, not universal constants cast in stone.

Finally, clarity of presentation is always important, particularly in this context, since the teams were instructed to play the role of consultants. This need for clarity placed a premium on having a concise, informative summary and on finding means of insightfully conveying the results. Verbal descriptions of algorithms and results are usually dry at best; but as the saying goes, a picture can be worth a thousand words. Teams that conveyed their results in well-designed tables or graphs typically made their point most effectively. All three of the outstanding teams distinguished themselves in this regard.

The Emergency Power-Restoration problem was difficult, and many excellent solutions were offered. In the end, however, three papers stood out from the others, and the members of those teams should feel proud of their accomplishments.

About the Author

Jonathan P. Caulkins did his undergraduate work at Washington University's School of Engineering and Applied Science, where he participated in the MCM. His teams' papers were rated Outstanding in the first two competitions, in 1985 and 1986.

Jon went on to earn a master's degree in electrical engineering and computer science and a doctorate in operations research from MIT. Now he is an assistant professor of operations research and public policy at Carnegie Mellon University's Heinz School of Public Policy and Management, where his research focuses on developing mathematical models of illicit drug markets.

Jon was an associate judge for the Emergency Power-Restoration Problem.

Author's Commentary:
The Outstanding Emergency
Power-Restoration Papers

Joseph Malkevitch
Dept. of Mathematics and Computing
York College (CUNY)
Jamaica, NY 11451

In the summer of 1991, hurricane Bob left considerable damage as it went up the East Coast of the United States. I live on Long Island, where Suffolk County was hit hard. Watching TV coverage of the havoc wrought by the hurricane gave me the idea for a mathematical modeling problem.

For me, teaching mathematical modeling offers the dually appealing features of being able to develop interesting mathematical theory in an environment that stresses both problem-solving and applications. Students who do not always share my esthetic for the delights of mathematics can rarely turn their backs on how mathematics is used, often in dramatic ways affecting daily life. Working on open-ended problems is exciting and fun, especially when in the social context of working with other people.

Modeling problems fall loosely into two types of questions: applied situations in which complete information is available about the phenomenon to be analyzed, and situations that lend themselves to analytic study but for which complete data are not available. The first type often has limited avenues of approach and leads to structured mathematical attack. However, I have a great fondness for the second type, where the very vagueness of the situation means that a variety of attacks suggest themselves, and that the "nature" of the solution depends on the attack used. Such problems encourage genuine "brainstorming," a process that yields great excitement and pleasure.

The power company responsible for restoring power to customers who have lost electricity seems to me to face hard choices. If the company begins efforts during the storm, it risks danger to its employees, plus the possibility that it might solve a problem in one area and shortly later have to dispatch another crew there for later damage. On the other hand, if the company waits until the storm is over, it runs the risk of adverse publicity due to customers who wait a long time until power is restored.

The power company has a variety of potential concerns to weigh, which include

- costs to restore power,

- time to restore power for politically sensitive sites, and

- time to restore power to all users.

Various tradeoffs seem possible in my mind, between costs to the power company (i.e., overtime; hiring extra crews; legal fees to deal with suits due to loss of life, food spoilage, or lost business opportunity) and costs to the power company's customers in time or money.

Another issue is how to locate the source of a power outage. When power is lost in an area, the company may learn of the outage from its own instruments but also from customers who call in. Some callers may have information of potential value to the company.

A key technical and policy question is determining a priority scheme for restoring normal functioning to the system. Once the power company has verified the causes of the outages, a dynamic system must be implemented to allocate crews with the proper equipment and training, get crews from one outage site to another, and keep personnel working up to par. Issues here include estimation of time to get between sites and tradeoffs between using regular personnel vs. extra crews especially hired to help.

It seemed to me that the situation of power restoration after a hurricane had a variety of tantalizing aspects, which could be made concrete enough to call forth solutions that could be compared in a reasonable manner, yet additional aspects would leave open a broad array of attack modes.

As I had fun playing with all of these aspects of the problem in my own mind, it gave me a healthy respect for what local power companies must do to be in a state of preparedness for power emergencies, large and small. Not only did I look forward to the creative ideas that contest participants might develop, but I also looked forward to learning more about what the power companies do in practice. Creating problems, it seems to me, is easily as much fun as working on their solutions!

Acknowledgment

I would like to thank Profs. Frank Giordano and Maynard Thompson for helping me to flesh out my original version of the problem.

About the Author

I was born and raised in Brooklyn, attended Stuyvesant High School and Queens College (CUNY), and received a Ph.D. in geometry under Don Crowe at the University of Wisconsin–Madison. My mathematical interests lie in geometry, mathematical modeling, the mathematics of equity, and mathematics education. I enjoy playing with my sons Alexander (20 mos.)

and Benjamin (2 yrs.) and listening to classical music, especially string quartets.

After I submitted the MCM problem, my understanding of the problem—and appreciation for solutions—was made personal when a windstorm knocked down a tree limb in our yard and severed a power cable, provoking a 12-hr power outage.

Restoring Power after Storm to Take Months and Millions

Catherine S. Manegold
Special to the *New York Times*
Saturday, September 5, 1992 (National Edition) 1, 6.
Copyright ©1992 by The New York Times Company.
Reprinted by permission.

MIAMI, Fla., Sept. 4 — The scope of the crisis that faced the Florida Power & Light Company after Hurricane Andrew [on Aug. 23–24] knocked out electricity to more than 1.4 million customers and demolished mile after mile of its delivery network is best grasped in its particulars.

There was the several-mile stretch of 152d Avenue near Princeton that was choked off by dozens of steel-reinforced concrete utility poles that had crashed directly across traffic lanes. Installed just three weeks before, the 65-foot, $5,000 poles broke clean at the base and toppled in the storm.

There were the vast stretches of U.S. Route 1 that sounded like a washboard under car tires as drivers rattled over downed power lines that littered the roadway from Coconut Grove to Florida City, 40 miles to the south.

Perhaps most telling of all, there was a humming sound at the utility's Princeton substation, which usually pumps out power to about 20,000 people in south Dade county. Until today, it was reduced to drawing power for its own lights, air conditioner and coffee makers not from the maze of transmission lines that crisscrossed the skies overhead, but from that new friend to South Florida: a grinding, portable generator that could power only a fraction of the substation's own needs.

The power company itself was without power for the first 30 hours after the storm; some company buildings were dark for 12 days.

Most residents could not grasp the extent of the damage. They knew only one thing about the power company's troubles: when they flipped a switch, nothing happened.

Stone-Age Conditions

Radio announcers have joked all week about south Dade county having been thrust back into the Stone Age. But in some ways the adjustment was even harder than it sounded. In the Stone Age, no one ever counted on electricity to transform a slab of meat into a meal. And generators did not malfunction and blow out refrigerators, television sets and other fragile impedimenta of modern life. Nor did people fume at the suddenly miserable interconnectedness of things, how everything—from the bank machines to the gas stations to the electric doors at local markets this week—responded to only one position: off.

For company officials focusing on the devastation south of Kendall Drive, the dimensions of the restart effort were overwhelming. "Basically, we have to rebuild a system that took us 50 years to put together," said Bob Marshall, the company's vice president for power distribution, speaking at the company headquarters after the lights went back on. "The only thing is, we've got to rebuild it in a matter of weeks."

Already, the company has restored about 90 percent of its service. But the last miles of this marathon will be hardest. Some devastated areas around Homestead have not even been surveyed for damage yet, though company officials are working on the assumption that destruction there was total.

Fragile, Makeshift Repairs

Today, more than 140,000 people in South Florida were still waiting for service. Some may have to wait two months or more.

Even then, all the usual safeguards and backup systems will not be in place for months, leaving the whole electrical system vulnerable to aggravating problems in even minor storms.

"Right now, we are building a fragile system," said Larry Taylor, vice president in charge of power delivery. He noted that South Florida's rainy season is just beginning and that power demands are at their peak, with temperatures routinely soaring into the 90's.

"It's like a house of cards," Mr. Taylor said. "A good lightning storm that would usually just leave you with a little flicker now will be able to black out a whole area."

The fierce summer squalls that stuck Florida last Saturday were a case in point, leaving some residents whose power had been restored in the dark for the second time in less than a week.

Getting the electricity back on is the power company's first priority, but it is hardly the only one; restoration of the entire system will not be completed for six months at least.

Once power is restored, workers must go back and examine bent poles and damaged equipment to see if they, too, must be repaired or replaced. "We've got one section that runs about six miles long where there is damage to every single pole," Mr. Taylor said. "That is 60 to 70 50-ton structures that are bent and twisted up. Some whole lines are all listing 5 to 10 degrees. There's miles of that. Some can stand for now, but we'll have to get to it later."

The cost of restoring service to the level of the day before Hurricane Andrew is not yet known. But it is clearly staggering.

$20 Million Deductible

"The only financial discussions that I have had so far have been about our insurance," James L. Broadhead, the utility's chief executive officer, said Thursday. "Other than that, so far we have just been in a war where the enemy is sending wave after wave of soldiers to the ramparts and we are just trying to beat them back. We just haven't had time to total anything up."

When the company does add up its losses, most will be probably be covered. Florida Power & Light is insured for $350 million against damage to its transmission and distribution systems. The damaged generating plants at Turkey Point and Cutler are covered for $6 billion. The utility's $20 million deductible will be paid from a $70 million storm fund kept for that purpose, Mr. Broadhead said.

Utilities' Largest Disaster

No crisis of this magnitude was ever planned for. None ever hit dead-on before. It was, by tallies of F.P.&L. officials, the largest natural disaster ever to challenge a modern utility company. The damage extended over three counties and included everything from wooden poles to transformers to the heavily damaged oil and nuclear plants at Turkey Point and oil-powered generators at Cutler along the coast.

Preliminary surveys indicate that the company will need almost 10,000 new transformers, at an average cost of $650 each, to replace those twisted, bent and battered in winds that reached 150 miles an hour. At least 18,000 wooden utility poles are already on order at costs that range from $140 to $1,000 each. Concrete poles take longer to manufacture, and will not be used in the first phases of the repair work. The ceramic sleeves that insulate the wires from the poles were shattered, cracked and broken. Nearly 180,000 of them, at $30 each, will be needed, too. Then the company needs more than 6.6 million feet of power lines, at 26 cents a foot. Another 2.6 million feet of lower grade wire will be needed for reconnecting houses to poles.

Repairs to the company's generating plants at Cutler and Turkey Point should take six months or more. C.O. Woody, a senior vice president in charge of power generation, said a 400-foot chimney at one of the Turkey Point oil plants was damaged beyond repair and would take six months and "several million dollars" to reconstruct. It was razed today in a controlled explosion. A second chimney also showed a hairline crack last week, and company officials were reviewing whether that chimney also need to be demolished and rebuilt. In the meantime, crews were starting to clean up a thin coat of oil that had spewed over the entire site when an oil tank ruptured at the height of the storm.

Anxiety About Nuclear Plant

The larger generating plants at Turkey Point, where the hurricane's eye passed directly overhead, suffered extensive damage, too. Though the two nuclear plants at the site survived with only "cuts and bruises," said Jerry Goldberg, the president of F.P.&L.'s nuclear division, they also are not expected to be back in operation for nearly six months as crews replace damaged security cameras, downed fences, battered warehouses and demolished offices and office equipment.

Restarting the complex will depend, too, on the emotional condition of the employees, Mr. Goldberg said. Explaining that about 80 percent of the utility's workers lost their homes in the storm, he said company officials wanted to be sure to take time to see that staff at the nuclear plant "could properly focus on their duties."

Though the nuclear plants at Turkey Point were shut down prior to the storm and escaped any significant damage, the hurricane did sent ripples of alarm through the nuclear industry. The storm knocked out all six communication links between the plant and the outside world, and for six or seven hours no one could get a message in or out. "We got one radio message out at about 7:30 that morning," said Mr. Goldberg, "but the silence caused a lot of apprehension in our offices and at the Nuclear Regulatory Commission."

Florida's weather, its geology and geography complicate an already complex task of reconstruction. The state gets more lightning than any other state in the nation, meaning more power surges and blackouts; the ground is a hard coral rock under a relatively thin layer of topsoil, making it difficult to drill holes for new poles; and because the state is a peninsula, supplies can come in only from the north, which causes traffic bottlenecks that slow relief supplies.

To make matters worse, the power company has had to deal with the same day-to-day privations, like closed gas stations, that have stretched everyone's patience.

"We've had 40 to 50 tractor trailers and at least 60 drivers working," said Dave Lindstrom, the company's manager of distribution and inventory services. "By mid-day Wednesday, we had delivery of over 12,000 poles alone. But all those trucks needed gas. And then the trucks in the field needed gas."

Of course, one reason there was no gas in south Dade county was because there was no power to pump it. So the company had to dispatch fuel trucks to Fort Everglades and leave drivers sitting in long lines, along with the police, the Red Cross and a variety of other emergency workers. "After waiting three hours or so to fill the tankers," said Mr. Lindstrom, "the drivers would go and get stuck in traffic for four hours."

By Friday, the company had more than 4,000 workers scouring the storm-ravaged area. But most of the crews were hired from neighboring states and from other Florida utility companies, and they needed a place to sleep.

At the Princeton substation, which is usually the work center for 63 crew members, last week there were more than 300 people working. Moving them about was a logistical challenge itself, said Ana Garcia, a customer relations agent who had been deputized as a group supervisor for the duration. She explained that crews must be bused in from area hotels starting at 5 A.M.

1993: The Coal-Tipple Operations Problem

The Aspen-Boulder Coal Company runs a loading facility consisting of a large coal tipple. When the coal trains arrive, they are loaded from the tipple. The standard coal train takes 3 hours to load, and the tipple's capacity is 1.5 standard trainloads of coal. Each day, the railroad sends three standard trains to the loading facility, and they arrive at any time between 5 A.M. and 8 P.M. local time. Each of the trains has three engines. If a train arrives and sits idle while waiting to be loaded, the railroad charges a special fee, called a *demurrage*. The fee is $5,000 per engine per hour. In addition, a high-capacity train arrives once a week every Thursday between 11 A.M. and 1 P.M. This special train has five engines and holds twice as much coal as a standard train. An empty tipple can be loaded directly from the mine to its capacity in six hours by a single loading crew. This crew (and its associated equipment) costs $9,000 per hour. A second crew can be called out to increase the loading rate by conducting an additional tipple-loading operation at the cost of $12,000 per hour. Because of safety requirements, during tipple loading no trains can be loaded. Whenever train loading is interrupted to load the tipple, demurrage charges are in effect.

The management of the Coal Company has asked you to determine the expected annual costs of this tipple's loading operations. Your analysis should include the following considerations:

- How often should the second crew be called out?

- What are the expected monthly demurrage costs?

- If the standard trains could be scheduled to arrive at precise times, what daily schedule would minimize loading costs?

- Would a third tipple-loading crew at $12,000 per hour reduce annual operations costs?

- Can this tipple support a fourth standard train every day?

Comments by the Contest Director

The problem idea was suggested by Gene Woolsey (Dept. of Mineral Economics, Colorado School of Mines, Golden, CO), based on his consulting for a coal company in Wyoming. The problem statement was formulated by Chris Arney and Jack Robertson (Dept. of Mathematical Sciences, U.S. Military Academy, West Point, NY).

Coal-Tipple Operations

John Petty
Ray Eason
Jon Rufenacht
U.S. Military Academy
West Point, NY 10996

Advisor: Charles G. Clark, Jr.

Solution Approach

We formulate a computer simulation in order to calculate operation cost projections. The simulation produces random weekly train schedules and evaluates these schedules with a Pascal program. The simulation has 10,000 weeks of iterations, and is supported by five submodels that aid in arriving at the information required by the Aspen-Boulder management.

Results

1. Annual cost projection $89,817,000
2. Monthly demurrage projection $3,053,000

Recommendations to Management

- We have formulated an ideal train schedule that minimizes demurrage costs. We recommend that this schedule be implemented in order to make the system more cost effective.

- In order to decrease demurrage costs, use two crews if a train must wait for the tipple to be filled.

- A third loading crew will reduce annual operating costs (cf. Submodel 2).

- This single tipple system can handle a fourth standard train every day. However, we caution that the system will "lag" on Thursdays.

- We recommend that our simulation be verified by comparison with data.

- We recommend that our simulation be used to explore the expected value of hypothetical scenarios.

- We recommend that our simulation be revised to model the operation more realistically.

The Model

Model Assumptions

- Upon arrival to the tipple system, all trains are completely empty.

- All train loading rates remain constant.

- The rate of tipple filling per crew remains constant, so adding a second crew doubles the overall rate.

- Assume that all events are discrete (i.e., trains can arrive only on the hour).

Model Formulation

In order to model the coal operation, we first performed a few hand calculations to get exactly how the operation behaved. After only a few trials, we discovered that many different arrival schedules are possible. It would not be feasible to enumerate every possible train arrival schedule by hand.

Another source of confusion is that the problem does not specify the exact arrival rate of the trains. Without an arrival rate, we were not able to construct easily an algebraic formula or linear program to model the behavior of the operation. Therefore, we decided upon the simulation approach. With the aid of an algorithm and a randomly generated train schedule, we could evaluate every possible scenario. The simulation assists in answering the questions of finding the expected annual costs of the tipple operation and the expected monthly demurrage costs. To assist with answering the other questions, we used five other submodels.

- Simulation (Main Model)

  - Task 1: Random Weekly Schedule Array
  - Task 2: Operation of System and Cost Evaluation

- Hand-calculated Submodels

  - Submodel 1: The Ideal Scenario
    * non-overlapping service times
    * no high-capacity train arrives
    * one crew fills tipple
  - Submodel 2: Minimizing Tipple Loading Costs
    * this is an algebraic optimization formula
  - Submodel 3: Minimizing Demurrage Costs

* look at multiple train arrivals
  – Submodel 4: Four Trains?
    * consider four standard trains arriving per day
  – Submodel 5: Worst-Case Scenario
    * look at maximized demurrage costs on a Thursday when all trains arrive at the same time

Simulation

The algorithm of the simulation will

* produce a random weekly train schedule;

* evaluate costs for this schedule;

* repeat step (1) for n iterations; and

* tabulate the expected weekly, monthly, and annual costs.

Tasks of the Simulation Model

* generate a random weekly train schedule, and

* evaluate weekly train schedule to model operation behavior and calculate costs.

Simulation Task 1: Random Weekly Schedule Array

Concept: Produce a random variable Y to slot train arrivals within the 168 hours of a week. Produce a random variable X to slot train arrivals of the high-capacity train.

Model formulation: In considering different modeling alternatives, we quickly decided upon a simulation model, because costs are based solely upon the probabilities of trains showing up at crucial times. To a great extent, once a train shows up, the system deals with that train in an extremely deterministic fashion. Since we have no underlying distribution, we found that a simulation model would probably give us the best estimate for an expected value of costs incurred.

Our first step was to determine the probabilities of trains showing up. We decided to look at each day in a global or outside-of-time viewpoint. If we tried to keep track of probabilities as we progressed through the hours of the day, we found that probabilities of train arrival from hour to hour changed, due to laws of conditional probability. However, when observed from the global viewpoint, in each 15-hr window of opportunity for a train to arrive each day, there was a 1-in-15 chance that a given train would show up within a given 1-hr window of time.

Since we decided to consider arrival probabilities on a train-by-train basis, we quickly decided that in evaluating the costs of one day's business, it would simplify matters greatly if we could extract all matters of probability to the beginning of our daily cost evaluations. Another consideration we had to take into account was that days of operation overlap, so that sometimes trains show up one day and don't leave until the next. Therefore, for purposes of observing costs over time, we needed to consider a longer period of time than just one day. For this reason, we extended our schedule length out to one week, or 168 hrs. In each hour, we needed to check to see if any trains arrive and then perform the necessary actions in reaction to the current state of the system. But since we wanted to extract the probabilities to the beginning of the model, we decided to generate ahead of time, independent of cost and time evaluations, a list of times that the trains arrive during a one-week time period.

For each train, this was a simple matter of generating a random number between 1 and 15 and "slotting" it into the appropriate window. For instance, for the first of three trains to arrive on day one of the week, if a 7 was generated, then a train arrived during the seventh hour after 0500, or between 1100 and 1200 hrs. For the purposes of our simulation, we simplified our model to say that any train that arrived during that time period would arrive at 1100 on the hour. This gave us a discrete set, from which we could run a discrete simulation. We continued this slotting until the entire arrival schedule for a week was completed.

Random Variables:

Y = a discrete uniform random variable on $0, \ldots, 14$

X = a continuous uniform random variable on $[0, 1]$

Assumptions:

- The distribution of arrival times is unknown.

- The random number generator of Turbo Pascal is sufficiently random.

Array Production Algorithm:

- Step 1: Generate a random number Y between 0 and 14 (15 hrs is the window of time between 5 A.M. and 8 P.M.).

- Step 2: Add 6 to Y. (We add 6 to Y because, relative to our timeline, a train cannot arrive before the 6th hour.)

- Step 3: Add a train to the number of trains generated for that hour. Designate a standard train with the value of 1 and a high-capacity train with the value of 10 (an arbitrary choice). We have "tagged" the hour block by manipulating $(6+Y)$. This "tagged" slot represents the hour that a train arrives. We must now add a value to this slot (1 for a standard train) to signify a train arrival.

- Step 4: Repeat the last three steps three times to slot three trains for that day.

- Steps 5–10: These steps simulate the slotting of the weekly train schedules for the rest of the 168-hr week (we save the high-capacity arrival for Step 11).

- Step 11: Generate a random number X between 0 and 1, add 84, and slot the arrival of the high-capacity train. (On our timeline, the high-capacity train can either arrive during the 84th or 85th hour: evaluate any arrival during the hour blocks at the beginning of the hour up to and including the end of the hour.)

Simulation Task 2: Operation of System and Cost Evaluation

Concept: Evaluate the weekly random schedule in a simulation that calculates labor and demurrage costs.

Model formulation: Once we determine our random arrival schedule, we have to evaluate the week of arrivals as it was generated. The algorithm proceeds hour by hour.

Simplifying Assumptions:

- If trains are waiting to be serviced and a high-capacity train arrives, then the high-capacity train will be loaded first because it has the highest demurrage cost.

- If no trains are waiting in line and no train is being serviced, then it is advantageous to fill the tipple with two crews.

- If you make a train wait because of an empty tipple, then it is more cost-effective to use two crews, in order to minimize demurrage costs (refer to Model 3 for a proof of this simplifying assumption).

Algorithm: (see **Figure 1**)

- Step 1: Check to see if any trains arrived at this hour. If so, how many trains are in the system, and is one of them the high-capacity train? If the high-capacity train arrives: Go to Step 7.

- Step 2: If three standard trains are in the system, at least two are going to have to wait on hold and incur demurrage costs. Add $30,000 to cost for this hour's waiting. Go to Step 5.

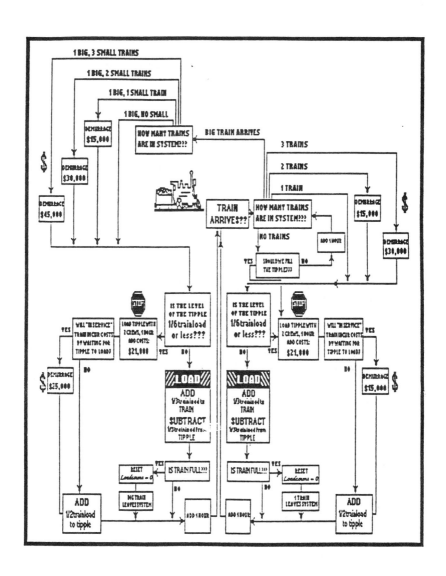

Figure 1. Algorithm for the model.

- Step 3: If two trains are in the system, at least one will have to wait and incur demurrage costs. Add $15,000 to costs for this hour's waiting. Go to Step 5.

- Step 4: If no trains are currently in the system, then let the system be idle, except check to see if the tipple is not at full capacity. If it is not at full capacity, then begin filling it. In 1 hr, two crews can fill one-half trainload into the tipple. Add $11,000 to cost for working crews if they fill the tipple during this hour. Go back and check for next hour to see if trains arrive. Proceed to next hour. Go to Step 1.

- Step 5: By now, we must see if we can actually fill the train that is on-line to be filled. If so, then take one-third of a trainload away from the tipple and put it into the train. If the train is full, send it on its way and reset to 0 the load value of the current train being loaded. Otherwise, keep track of the size of the load in the current train being loaded. Proceed to the next hour. Go to Step 1.

- Step 6: If the tipple needs filling, we should fill it with two crews, add a half trainload to the tipple, add $1,000 demurrage cost for this train to sit on the tracks waiting for the tipple to load, and add $21,000 to the labor costs for the crews at work. Proceed to next hour. Go to Step 1.

- Step 7: Now that a high-capacity train with five engines has arrived, to prevent exorbitant demurrage costs for making this train wait in lieu of another, we will automatically switch to service the high-capacity train. If another train is currently being serviced, then we will put it on hold for now.

- Step 8: If three standard trains are now waiting, all three are going to have to wait for the high-capacity train and incur demurrage costs. Add $45,000 to costs for this hour's waiting.

- Step 9: If two standard trains are now waiting for the high-capacity train to be serviced, both will have to wait and incur demurrage costs. Add $30,000 to costs for this hour's waiting.

- Step 10: If one standard train is now waiting for the high-capacity train to be serviced, then the standard train will have to incur demurrage costs for this hour. Add $15,000 to costs for this hour of waiting.

- Step 11: By now, we must see if we can actually fill the high-capacity train that is on-line to be filled. If so, then no demurrage costs will be incurred for this one train for this hour. If the tipple needs filling, we should fill it with two crews, add a half trainload to the tipple, add to costs $25,000 for the high-capacity train to sit on the tracks waiting and $21,000 for the crews at work. Proceed to the next hour and go back to Step 1 now if the tipple was filled and the high-capacity train had to wait.

- Step 12: If we did not need to fill the tipple, then take one-third of a trainload away from the tipple and put it into the train. If the high-capacity train is full, send it on its way and let the load value of the current train being worked on go to 0 since it has now left. Otherwise, keep track of the size of the load in the current train being serviced. Proceed to next hour and go back to Step 1.

Figure 1 gives a graphic representation of the algorithm. Note that when we check to see if the tipple needs refilling, we check to see if it is less than one-sixth full. This is simply because of the nature of the fractions at which the tipple empties itself and is refilled hour by hour. By setting our refill value at one-sixth or less, we prevent our simulation from accidentally dropping the tipple load value below zero.

Results: Results from the simulation are given in **Table 1**. Two trials may seem insufficient, but realize that each trial models 10,000 weeks of data, for a total of more than 384 years.

Table 1.
Results of simulation. Each trial is for 10,000 weeks.

	First trial	Second trial
Demurrage costs/yr	39,752,000	39,605,000
(Labor + demurrage)/yr	89,904,000	89,730,000

Error Analysis:

- Sources of Possible Error

    - oversimplification of coal operation
    - assumption of constant tipple fill rates
    - assumption of constant train fill rates

- Computer roundoff error. Only addition was used to tabulate cumulative costs. Since multiplication tends to magnify round-off error, we avoided this source of error. In the final calculation of costs per week, however, we used the formula

$$\frac{\text{total costs of } N \text{ iterations of weeks}}{N \text{ weeks}}.$$

If N is on the order of 10,000 and the total cost on the order of magnitude of 10^8, we may lose a few digits of accuracy.

Weaknesses and Strengths

Weaknesses of the Model

- There is no comparison to a real-world data set. The final step of the modeling process should be to verify the model against real-world data. For our problem, real-world data are not available for comparison. This is a major weakness.

- The assumption of discrete hours and discrete events limits the flexibility of the model. For example, if we could calculate demurrage costs for trains arriving at 1:15 and waiting until 1:35, the simulation cost projections would be more realistic.

- The assumption that trains arrive with empty payloads fails to consider that a train may arrive with a fraction of a payload already loaded.

- The assumption that workers are on call at all times fails to account for real-life situations. Can we depend on having two crews available regardless of the time of day, weather conditions, or holidays? Realistically speaking, we cannot. Our model, however, assumes that we can.

- We fail to take into account "hidden" costs. What happens to the operation if the tipple breaks down? What effects does depreciation have on the value of the equipment used in the operation?

Strengths of the Model

- With an 80386 personal computer, a 10,000-week iteration model can be run in less than 10 minutes.

- Our simulation takes a rather complex scenario and simplifies the coal operation into something that is manageable and that we can use to make predictions about the behavior of the system. Simplicity can be a powerful tool for understanding a complex world.

The Submodels

Submodel 1: The Ideal Scenario

If the standard trains could be scheduled to arrive at precise times, what daily schedule would minimize loading costs?

Our simplifying assumptions are:

- non-overlapping service times,

- no high-capacity train arrives,

- one crew fills tipple, and

- the day of the week is not Thursday.

We decided to start with the simplest of cases. This train schedule attempts to fit all of the trains within the window of 5 A.M. to 8 P.M. without multiple trains arriving at the same time. In addition, the one train that is being serviced at the tipple must remain until it is full without having another train arrive. After the train is full, the tipple must be refilled to the level of one trainload before another train arrives. When the next train comes, it will also remain until it's full without another train arrival. Again the tipple will be filled to the level of one trainload before the last train arrives. After all three trains have come and gone, refill the tipple to 1.5 trainloads. We have formulated this scenario as the ideal case. It is ideal because no train ever has to wait and therefore, no demurrage costs are incurred. The labor costs for the tipple loading crew (one crew at $9,000/hr, not two crews at $21,000/hr) are also minimized. If the Aspen Company has control over when trains can arrive and it is feasible to devise the most ideal (minimized costs) scenario, then this simple model satisfies these needs. One ideal schedule that works is listed in **Table 2**.

Table 2.
An ideal schedule of train arrivals.

Time		Tipple level	Fill tipple?
0500	Train A arrives	1.5	—
0800		0.5	Fill 0.5
1000	Train B arrives	1.0	—
1300		0.0	Fill 1.0
1700	Train C arrives	1.0	—
2000		0.0	—
0200		1.5	—

In the ideal schedule, there is one 5-hr gap between trains (3 hrs to fill the train, 2 hrs to fill the tipple with half a trainload) and one 7-hr gap between trains (3 hrs to fill the train, 4 hrs to fill the tipple with one trainload). Therefore, there are two ways for this schedule to work:

- The first time gap between trains A and B is 5 hrs (a 7-hr gap must follow between trains B and C to fill the tipple to one trainload).

- The first gap between trains A and B is 7 hrs (a 5-hr gap must then follow between trains B and C to fill tipple to one trainload).

Table 3.

Two possible ideal train schedules.

	A	B	C
Case a	0500–0800	1000–1300	1700–2000
Case b	0500–0800	1200–1500	1700–2000

Observe these two possible cases in **Table 3**.

In this manner, we can follow the coal operation in an ideal day (no overlapping service times and no demurrage costs). See **Table 4**. For such a day, the daily loading costs are $108,000, with no demurrage costs.

Table 4.

Minimum loading costs on a non-Thursday. Total costs: $108,000.

Time	Tipple (trainloads)	Train A	Train B	Train C	Loading crew
500	1.50				
600	1.50	arrives			
700	1.16				
800	0.83				
900	0.50	leaves			
1000	0.75				$9,000
1100	1.00		arrives		$9,000
1200	0.67				
1300	0.33				
1400	0		leaves		
1500	0.25				$9,000
1600	0.50				$9,000
1700	0.75				$9,000
1800	1.00			arrives	$9,000
1900	0.67				
2000	0.33				
2100	0			leaves	
2200	0.25				$9,000
2300	0.50				$9,000
2400	0.75				$9,000
100	1.00				$9,000
200	1.25				$9,000
300	1.50				$9,000
400	1.50				

We must not forget to include an analysis of the ideal scenario on Thursday. On Thursday, however, there is no way to avoid demurrage costs. The daily loading costs for an ideal Thursday are $210,000, with daily demurrage cost of $135,000, for a total daily cost of $345,000. See **Table 5**.

Table 5.
Minimum loading costs on a Thursday. Total costs: $345,000

Time	Tipple (trainloads)	Train A	Train B	Train C	Big Train	Loading crew
500	1.50	arrives				
600	1.16					
700	0.83					
800	0.50	leaves				$21,000
900	1.00		arrives			
1000	0.67					
1100	0.33					
1200	0		leaves			$21,000
1300	0.50				arrives	
1400	0.16					
1430	0					
1500	0.25				$25,000	$21,000
1600	0.75				$25,000	$21,000
1700	1.25				$25,000	$21,000
1730	1.50					
1800	1.33					
1900	1.00					
2000	0.67			arrives		
2100	0.33			$15,000		
2200	0			$15,000	leaves	$21,000
2300	0.50			$15,000		$21,000
2400	1.00			$15,000		
100	0.67					
200	0.33					
300	0			leaves		$21,000
400	0.50					$21,000
500	1.00					$21,000
600	1.50					

Submodel 2: Minimizing Tipple Loading Costs

Would a third tipple-loading crew at $12,000/hr reduce annual operations costs?

We derive an algebraic formula to model the cost for multiple crews. For notation we will use

Cost   is the total cost of labor,

P is the percentage of the tipple to be filled,

c is the number of crews, and

a is the number of trains in the system.

Our approach is to define cost as a function of the number of crews, the number of trains in the system, and the percentage of the tipple to be filled:

$$\text{Cost} = f(P, c, a).$$

If it takes any one crew 6 hrs to fill the tipple, and all crews work at this standard rate, we know that it takes $6/c$ hrs to fill the tipple. The cost of the tipple loading crew(s) is given by $12{,}000c - 3{,}000$. For 1, 2, and 3 crews, the costs are \$9,000, \$21,000, and \$33,000.

$$
\begin{aligned}
\text{Cost} &= P \times \text{hours} \times (\text{labor} + \text{demurrage}) \\
&= P \times \frac{6}{c} \times ([12{,}000c - 3{,}000] + 15{,}000a) \\
&= P \left(72{,}000 + \frac{90{,}000a - 18{,}000}{c} \right).
\end{aligned}
$$

For any integer $a > 0$, cost will be minimized by increasing c. Even though the second crew costs \$12,000/hr, or \$3,000 more than the first crew, the two get the job done in half the time. Demurrage savings of around \$15,000 per train per hour, minus an increase of wages of \$3,000, is well worth the extra crew.

Now let us concentrate on the question: "Would a *third* tipple-loading crew at \$12,000/hr reduce annual operations costs?" Yes! Substitute $c = 3$ into the above formula. Provided the third crew is called in whenever $a > 0$ (a is an integer), the third crew will always reduce that cost. Therefore the annual cost will be reduced also.

Submodel 3: Minimizing Demurrage Costs

How often should the second crew be called out?

Our simplifying assumption is that there are multiple train arrivals for any given hour.

Our reply to the question is that if there is at least one train waiting to be filled, it is cheaper to have two crews working to get the job done in half the time.

What would happen if the trains had to wait for half of a trainload to be filled into the tipple? For one crew, the demurrage cost would be 2 hrs times \$15,000, or \$30,000. For two crews, there would be a demurrage cost of 1 hr times \$15,000. Even for only half a trainload, the increase in demurrage cost is $15,000 vs. an increase of $12,000 labor cost. Using two crews would save $3,000.

Submodel 4: A Fourth Train?

Can this tipple support a fourth standard train every day?

Adding in a fourth standard train on a non-Thursday is possible, and it will not overload the system.

On a Thursday, however, the system will begin to get backed up. Here's why: Looking at the best possible scenario, two trains arrive at 8 P.M., and the tipple is empty until it finishes with the big train at 10 P.M. Not until 5 A.M. the next morning does the last train even begin to get filled. Assuming that we can schedule when the trains arrive, by Saturday evening the system can be back in line again. Therefore, since the system does not overload, it can handle a fourth standard train. Nevertheless, having four standard trains on Thursday will greatly increase the demurrage cost. [EDITOR'S NOTE: For space reasons, we omit the authors' schedules for four standard trains on Thursday, Friday, Saturday, and Sunday.]

Submodel 5: Worst-Case Scenario

Given that our simulation model produces expected costs per week, how can we "double-check" the results to see if the simulation output is reasonable?

We put tremendous effort into developing a flowchart and computer code of the coal operation, so human error is great concern. Even with subroutine checks and debugging, we felt that we needed a method of gauging a "ballpark" figure of expected weekly costs. Therefore, we investigated maximized costs scenarios. From this vantage point, we get some idea of the simulation model verification. [EDITOR'S NOTE: For space reasons, we omit the authors' schedules that justify the following worst-case costs.]

The two scenarios are:

- Three standard trains arrive together:

Labor Costs	$117,000
Demurrage Costs	$195,000
Total Costs	$312,000

- Three standard trains and a high-capacity train arrive together on Thursday:

Labor Costs	$210,000
Demurrage Costs	$655,000
Total Costs	$865,000

Conclusions and Recommendations

Using our simulation and submodels, we were able to answer successfully the questions listed below. We recommend to the management that our simulation model be employed to answer "what if" questions about expected value. Furthermore, we recommend that a third crew be added full time to reduce costs of demurrage. Below are the bottom-line answers to the questions posed by management.

- *What is expected annual cost of the tipple's loading operation?*

Results from first 10,000 week run	$89,904,000
Results from second 10,000 week run	$89,730,000
Average	$89,817,000

- *How often should the second crew be called out?*

  If there is at least one train waiting to be filled, it is cheaper to have two crews working, according to Submodel 3.

- *What are the expected monthly demurrage costs?*

Results from first 10,000 week run	$3,058,000
Results from second 10,000 week run	$3,047,000
Average	$3,053,000

- *If the standard trains could be scheduled to arrive at precise times, what daily schedule would minimize loading costs?*

  See **Tables 4** and **5**.

- *Would a third tipple-loading crew at $12,000/hr reduce annual operations costs?*

  Yes, according to Submodel 2.

- *Can this tipple support a fourth standard train every day?*

  Yes, according to Submodel 4.

References

Giordano, Frank R., and Maurice D. Weir. 1985. *A First Course in Mathematical Modeling*. Monterey, CA: Brooks/Cole. After we had decided on a simulation approach, we referred to the Harbor System Model (pp. 280–289) to see if we were on the right track; that model is similar to the coal-tipple problem.

Practitioner's Commentary: The Outstanding Coal-Tipple Operations Papers

Ruth Maurer
Dept. of Mathematical and Computer Sciences
Colorado School of Mines
Golden, CO 80401

Introduction

For the system of two crews, three regular trains, and one special train, all three of the final papers generated the same total cost (in the $87–90 million range, which is good for this stochastic situation). However, one team considered only a five-day week, while the others used a seven-day week.

Similar schedules for the three regular trains were achieved in the case where those trains could be scheduled. Costs ranged from $52 million to $59 million, again similar results given the stochastic nature of the problem.

All agreed that use a of third crew would reduce total annual costs but disagreed significantly on the total amount of savings.

All agreed that the system could handle a fourth regular train, but Thursdays would be problematic and costs may soar.

Detailed Analysis

The teams will be discussed in order of performance, first to last.

The team from Cornell University had the simplest solution and the one most amenable to sensitivity analysis. This team used an existing simulation package to build its model, and the parameters of the model can easily be changed to ask "What if —?" types of questions. This team's presentation of approach and results is probably the clearest, except for the the statistical analysis, which is not clear.

The team from the U.S. Military Academy wrote the clearest summary of the problem, approach, and recommendations. They wrote the most thorough statement of its algorithm, having done their own programming in Pascal; they also used spreadsheets to advantage in summarizing results. I disagree with their assumption of train arrivals on the hour—this is just unrealistic. They state as an assumption that "the distribution of arrival times is unknown" when they are actually assuming a discrete uniform distribution; they could just as easily have used a continuous uniform distribution.

The team from the University of Alaska Fairbanks is clearly more oriented in the direction of mathematical statistics, and less in the direction of simulation, than the other teams. This team's solution considers only five scenarios or "rule sets" and chooses the one that minimizes cost. Their solution to the scheduled-regular-trains part of the problem is reasonable, but they don't tell us how they arrived at that particular schedule. Their answer to the four-regular-trains question is sketchy at best. The team is to be commended, however, for testing their primary results against those of a Pascal simulation program (which they wrote). Since the simulation was based on the same logic as the theoretical solution, however, one would expect the results to agree.

About the Author

Dr. Ruth Maurer is presently Associate Professor of Mathematics (Operations Research/Applied Statistics) at the Colorado School of Mines. In addition to considerable professional work as a consultant, she is former Mayor of the city of Golden, Colorado. She also was the Consulting Energy Economist for the First Interstate Bank of Denver and was visiting professor of engineering at the U.S. Military Academy at West Point. For her pro bono consulting work for the Department of the Army, she was awarded the Outstanding Civilian Service Medal and the Commander's Medal. She is the co-author (with R.E.D. Woolsey) of the five books in the Useful Management Series.

Judge's Commentary: The Outstanding Coal-Tipple Operations Papers

Jonathan P. Caulkins
Heinz School of Public Policy and Management
Carnegie Mellon University
Pittsburgh, PA 15213–3890

This problem is deceivingly difficult because it has characteristics that are familiar (queueing, inventory, scheduling) but does not fit neatly into any standard class of problems. As a result, even though teams tried methods as varied as linear programming and simulated annealing, essentially every team also used Monte Carlo simulation.

Unfortunately, many teams plunged too quickly into simulation and neglected to perform supporting analyses. Indeed, several teams used Monte Carlo methods to estimate quantities, such as the expected value of the minimum of independent uniform random variables, that can easily be found exactly. These papers, even when the simulations were well constructed, typically generated more numbers than insight.

The better papers, including all three Outstanding papers, augmented simulations with other analysis. The U.S. Military Academy team produced upper and lower bounds on costs to verify that the results of its simulation were reasonable. When Thursday's operations spill over into Friday, demurrage costs on Friday may be affected. However, exact analysis of the costs of such disruptions is exceedingly difficult, and most teams simply ignored them. The University of Alaska Fairbanks team, in contrast, attempted to construct lower and upper bounds for those costs.

Better papers were distinguished also by a more mature treatment of the assumptions, including performing sensitivity analysis with respect to those assumptions. The Cornell University team stood out because they did not jump to the conclusion that two crews can fill the coal tipple twice as quickly as one. Using a simple graph, the team showed how demurrage costs vary with how much a second crew speeds up the loading process.

The three outstanding papers did not monopolize insightful analysis. For example, several teams gave careful discussions of whether or not to give priority to a high-capacity train. One even considered preemptive priorities and concluded that a standard train should preempt a high capacity train when the high-capacity train is less than about 20% full. There were also several excellent analyses of when to use one crew or two to refill the tipple, as a function of the amount of coal in the tipple, the time of day, and the number of trains still to come that day.

The questions of priority and number of crews were approached at different levels of sophistication. Some teams recommended a reasonable course of action with little explanation. Such recommendations are of little value, because they must be taken on faith and do not bring out general properties. Better papers gave either insightful intuitive arguments that produced more understanding or else mathematical proofs that were more persuasive. The best papers gave both.

More generally, good papers had a sense of perspective. Weaker papers worried excessively about minutiae, such as computer roundoff error and the fact that random number generators do not produce truly random numbers. Better papers recognized the general structure of the problem and designed their approach around it (Cornell); neatly presented and evaluated alternative decision rules (University of Alaska Fairbanks); and distilled their results into a concise, well-written summary (U.S. Military Academy).

When addressing a real-world modeling problem, there is no such thing as having "finished" the problem: There are always more ways of interpreting, structuring, and approaching the problem. In fact, that is the value of publishing the Outstanding papers—reading them helps provoke contestants to think in new ways about the problem that they worked on. Several ideas that were not fully tried in any paper but occurred to the judges as potentially fruitful include explicitly defining a state-space description of the system, and using dynamic programming or Markov decision processes, either as a solution procedure or as a way of structuring the problem.

The judges rewarded teams who used multiple methods, conducted sensitivity analyses, derived and justified insightful properties of good solutions, and had the perspective to understand and acknowledge the limitations of their models. It is very difficult to do well in all of these dimensions; good modeling is an art that takes considerable skill and practice for proficiency. The judges are delighted that so many students have accepted the challenge to become good modelers.

About the Author

Jonathan P. Caulkins participated in the MCM at Washington University, where his teams' entries were judged Outstanding in the first two competitions. Jon earned a master's degree in electrical engineering and computer science and a doctorate in operations research from MIT. Now he is an assistant professor of operations research and public policy at Carnegie Mellon University's Heinz School of Public Policy and Management, where his research focuses on mathematical models of illicit drug markets. Jon was an associate judge for the Coal-Tipple Operations Problem.

1994: The Communications Network Problem

In your company, information is shared among departments on a daily basis. This information includes the previous day's sales statistics and current production guidance. It is important to get this information out as quickly as possible.

Suppose that a communications network is to be used to transfer blocks of data (files) from one computer to another. As an example, consider the graph model in **Figure 1**.

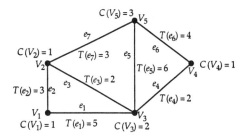

Figure 1. Example of a file transfer network.

Vertices $V_1, V_2, \ldots, V_m$ represent computers, and edges $e_1, e_2, \ldots, e_n$ represent files to be transferred (between computers represented by edge endpoints). $T(e_x)$ is the time that it takes to transfer file e_x, and $C(V_y)$ is the capacity of the computer represented by V_y to transfer files simultaneously. A file transfer involves the engagement of both computers for the entire time that it takes to transfer the file. For example, $C(V_y) = 1$ means that computer V_y can be involved in only one transfer at a time.

We are interested in scheduling the transfers in an optimal way, to minimize the total time that it takes to complete them all. This minimum total time is called the *makespan*. Consider the three following situations for your company:

Situation A

Your corporation has 28 departments. Each department has a computer, each of which is represented by a vertex in **Figure 2**. Each day, 27 files must be transferred, represented by the edges in **Figure 2**. For this network, $T(e_x) = 1$ and $C(V_y) = 1$ for all x and y. Find an optimal schedule and the makespan for the given network. Can you prove to your supervisor that your makespan is the smallest possible (optimal) for the given network? Describe your approach to solving the problem. Does your approach work for the general case, that is, where $T(e_x)$, $C(V_y)$, and the graph structure are arbitrary?

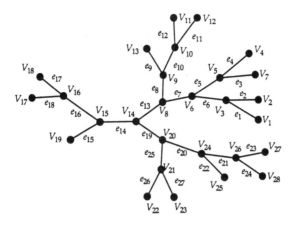

Figure 2. Network for situations A and B.

Situation B

Suppose that your company changes the requirements for data transfer. You must now consider the same basic network structure (again, see **Figure 2**) with different types and sizes of files. These files take the amount of time to transfer indicated in **Table 2** by the $T(e_x)$ terms for each edge. We still have $C(V_y) = 1$ for all y. Find an optimal schedule and the makespan for the new network. Can you prove that your makespan is the smallest possible for the new network? Describe your approach to solving this problem. Does your approach work for the general case? Comment on any peculiar or unexpected results.

Table 2.

File transfer time data for situation B.

x	1	2	3	4	5	6	7	8	9	10	11	12	13	14
$T(e_x)$	3.0	4.1	4.0	7.0	1.0	8.0	3.2	2.4	5.0	8.0	1.0	4.4	9.0	3.2

x	15	16	17	18	19	20	21	22	23	24	25	26	27
$T(e_x)$	2.1	8.0	3.6	4.5	7.0	7.0	9.0	4.2	4.4	5.0	7.0	9.0	1.2

Situation C

Your corporation is considering expansion. If that happens, there are several new files (edges) that will need to be transferred daily. This expansion will also include an upgrade of the computer system. Some of the 28 departments will get new computers that can handle more than one transfer at a time. All of these changes are indicated in **Figure 3** and **Tables 3–4**. What is the best schedule and makespan that you can find? Can you prove that your makespan is the smallest possible for this network? Describe your approach to solving the problem. Comment on any peculiar or unexpected results.

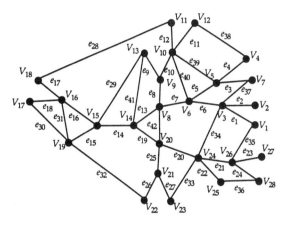

Figure 3. Network for situation C.

Table 3.

File transfer time data for situation C, for the added transfers.

x	28	29	30	31	32	33	34	35	36	37	38	39	40	41	42
$T(e_x)$	6.0	1.1	5.2	4.1	4.0	7.0	2.4	9.0	3.7	6.3	6.6	5.1	7.1	3.0	6.1

Table 4.

Computer capacity data for situation C.

y	1	2	3	4	5	6	7	8	9	10	11	12	13	14
$C(V_y)$	2	2	1	1	1	1	1	1	2	3	1	1	1	2

y	15	16	17	18	19	20	21	22	23	24	25	26	27	28
$C(V_y)$	1	2	1	1	1	1	1	2	1	1	1	2	1	1

Comments by the Contest Director

The problem was contributed by Joe Malkevitch (Dept. of Mathematics and Computer Science, York College (CUNY), Jamaica, NY) and Steve Horton (Dept. of Mathematical Sciences, U.S. Military Academy, West Point, NY).

Talking Fast: Finding the Makespan of a Communications Network

Matthew Prochazka
Wendy Schaub
Gary Amende
Beloit College
Beloit, WI 53511–5595

Advisor: Philip D. Straffin

Summary

We were asked to find an optimal schedule and makespan for three different situations, as well as to provide a generalized algorithm for the optimal schedule and makespan.

To check all the possible schedules is very inefficient and time-consuming and, on a large network, could take several days or weeks. Instead, we implemented a greedy algorithm, based on the expectation that optimizing locally will result in global optimization.

The makespan will be at least as long as it takes for any one computer in its network to finish its data transfers. So the computer that takes the longest to finish its data transfers should start transferring first. By choosing the best possible transfer at each step in the algorithm, the schedule was progressively built up into a prediction for an optimal schedule and makespan.

We found optimal schedules and makespans (3, 23.0, and 29.6) for situations A, B, and C. We also proved the correctness of these makespans found by our algorithm. Our method can be applied to a general network; although our computer program does not guaranteed optimality, a near-optimal schedule and makespan are produced.

For situations A and B, we recommend implementing the optimal schedules. For situation C, we recommend switching some computers capable of transferring more than one file at a time to locations that require multiple file transfers. In situation C, if each computer transfers only one file at a time, the makespan remains the same; but if the transfer times do not change when computers are switched, the makespan can be improved by 35%.

Assumptions

- Each network has computers, represented by vertices $(V_1, V_2, \ldots, V_m)$ and transfers between them, represented by edges $(e_1, e_2, \ldots, e_n)$. Each computer has a designated capacity, $C(V_i)$, to transfer files simultaneously, and each transfer has a designated time necessary to complete, $T(e_j)$.

- The computers do not break down and are always functioning.

- All computers in the network must receive the information on a daily basis.

- Equal priority is placed on all files.

- All transfers are made directly, not via a third party. Thus, all computers have the necessary information to make the transfer. This assumption avoids the problem of waiting for information from another computer before a transfer can be made (which would obviously increase the makespan).

- The process of transferring files does not have to wait for the implementation of this model. This assumption avoids any delays that depend on the running time of the computer model.

- Given a schedule, the company knows how to implement it.

Analysis of the Problem

The minimum time required by each computer can be calculated using its capacity and the number of connecting transfers. For example, computer V_{20} in **Figure 2** of the problem statement must transfer three files, each of which requires seven time units to complete. Since the computer can transfer only one file at a time, the file transfers must be done sequentially, producing a minimum total time of 21 time units. In this way, a minimum time can be found for each computer in the network.

The makespan for this network will be at least the largest of these minima, since each computer must complete all of its transfers. A starting point would be to initiate transfers that require the maximum time to complete. Once those computers are transferring data, the constraints of the network indicate which others can begin transferring at the same time. When a transfer is complete, two computers become available to work on other transfers. At this time, a re-evaluation of the network is needed to determine if another transfer can be initiated. By re-evaluating the network every time a transfer is completed, the idle time is reduced to a minimum. In this way, a minimum overall transfer time can be achieved.

Model Design

We decided to solve this problem using a greedy algorithm. This is an optimization algorithm that applies a simple-minded strategy of progressively building up a solution, one element at a time, by choosing the best possible element at each iteration. Greedy algorithms are based on the hope that optimizing locally (at each step, we do as well as we can) will result also in global optimization (at the end, we have an optimal solution overall). In our case, our elements were computers and we were trying to choose the best possible transfers to initiate at each iteration to schedule the makespan.

We first had to arrange the computers in a priority list. Doing so allowed us to choose the two best possible computers to transfer data between at each iteration. We used a modified bin-packing algorithm, which calculates the minimum time needed for a computer to finish (complete its transfers).

Modified Bin-Packing Algorithm: For each computer V_i, let the number of bins be $c = C(V_i)$. Each V_i must perform a subset of the set of all transfers in the network.

1. Suppose that V_i must perform k transfers. Rename these transfers $e_{(1)}, \ldots, e_{(k)}$ so that $T(e_{(1)}) \leq T(e_{(2)}) \leq \cdots \leq T(e_{(k)})$.

2. If $k \leq C(V_i)$, place one transfer in each bin. Thus, $T(e_{(k)})$ is the minimum time for that computer to finish.

3. If $k > C(V_i)$, place $T(e_{(k)})$ in the first bin, $T(e_{(k-1)})$ in the second, and so on, until all the bins contain one transfer. Place the next longest transfer, $T(e_{(k-c-1)})$, in the bin containing the lowest transfer time, $T(e_{(k-c)})$, and sum their two times. Place the next longest transfer, $T(e_{(k-c-2)})$ in the bin containing the lowest summed transfer time, and sum their times. Continue in this way until all transfers are placed in bins. The bin with the greatest summed transfer time is the minimum time for that computer to finish.

Using this algorithm, the computer that requires the most time to finish gives a lower bound for the entire network transfer time.

By placing the minimum transfer times needed by each computer in ascending order, we obtain a priority list; and the greedy algorithm can be used to find a minimal schedule. To initiate transfers, the two computers with the longest minimum transfer times are connected if a connection is possible. If no connection exists between these two computers, a connection to the next computer on the priority list is attempted. This continues until a successful connection is made.

At this stage, there are two possibilities for further connections: Either the capacity of the first computer on the priority list is equal to 1, or it is greater than 1. In the case where the capacity equals 1, the computer is no longer available to make other transfers. Therefore, the first available computer on the priority list is connected to the next computer on the list to which a connection is possible. This continues until all possible connections are made.

If the capacity is greater than 1 for the computer requiring the greatest transfer time, the computer is available for other transfers. A connection is then attempted between this computer and the next available computer on the priority list, until either the number of connections made by the first computer equals its capacity or no other connections are possible. Therefore, all possible transfers involving this computer are in progress. This procedure is repeated for each computer, working in descending order of the priority list until all possible connections are made and a minimal schedule achieved.

Computer Implementation

We developed a Pascal computer program to enable the company to analyze their computer network and obtain a minimal schedule. The program produces the minimum time for each computer to finish, calculates a schedule of times to start each of the transfers, and reports the total elapsed time of the minimal schedule. This program can be analyzed using big-oh notation [Sedgwick 1988]:

Definition. *A function $g(N)$ is said to be $O(f(N))$ if there exist constants c_o and N_o such that $g(N) < c_o f(N)$ for all $N > N_o$.*

For a problem of a certain size, the O-notation is a useful way to state upper bounds on running time that are independent of both inputs and implementation details. In our case, the upper bounds are independent of computer capacities and transfer times between computers. We found that our program has a running time of $O(m^3)$ for m computers. Closer examination shows that in the worst-case scenario, the constant c_o is close to $1/4$; in the average case, however, this constant is much smaller.

Situation A

The corporation has 28 computers, each with capacity 1, and 27 transfers, each with transfer time of 1 unit. With our algorithm, we found the makespan to be 3 units. The makespan cannot be any lower than this, because any computer on the interior of the network requires at least 3 units to complete its transfers. There is more than one way to optimally schedule the transfers, such as the schedule shown in **Table 1**.

Table 1.

Schedule for situation A.

At time	Start transfers
0.0	$e_2, e_5, e_{10}, e_{13}, e_{16}, e_{21}, e_{25}$
1.0	$e_1, e_3, e_7, e_9, e_{11}, e_{14}, e_{17}, e_{20}, e_{23}, e_{27}$
2.0	$e_4, e_6, e_8, e_{12}, e_{15}, e_{18}, e_{19}, e_{22}, e_{24}, e_{26}$

Total time elapsed: 3.0 units

Situation B

The network is the same as in situation A, but the transfer times are different. The maximum time required by any of the computers is 21 units, which is the minimum time required by computer V_{20} to finish. Thus, a lower bound of the makespan for this network is 21 units. Our program, however, predicts a makespan of 23 units, as seen in **Table 2**.

Table 2.

Schedule for situation B.

At time	Start transfers
0.0	$e_4, e_6, e_{10}, e_{13}, e_{16}, e_{20}, e_{24}, e_{26}$
5.0	e_{23}
7.0	e_3, e_{22}
8.0	$e_2, e_9, e_{12}, e_{15}, e_{18}$
9.0	e_7, e_{19}, e_{27}
11.2	e_{21}
12.1	e_1
12.2	e_5
12.4	e_{11}
12.5	e_{17}
13.0	e_8
16.0	e_{14}, e_{25}

Total time elapsed: 23.0 units

The discrepancy can be explained by observing the architecture of the network. Computer V_{20} requires a minimum of 21 units to finish (e_{19}, e_{20}, e_{25}), because each transfer requires 7 units and the computer can transfer only one file at a time. This is the maximum time required by any of the computers in the network, so it is a lower bound.

To achieve this lower bound, the first transfer should include V_{20}. Looking at all the possible transfers involved with V_{20}, one could start with e_{20}. When e_{20} finishes, e_{19} or e_{25} should be initiated immediately, and thus e_{13} and e_{26} cannot both be transferring at this time. If e_{13} is not transferring, e_{19} would be initiated, which takes 7 units to complete. Now, e_{13} cannot begin transferring until these 7 units have elapsed. So after 14 units (7 from e_{20} and 7 from e_{19}), e_{13} can begin. Transfer e_{13} takes 9 units to complete; so by the time that it is done, 23 units will have elapsed. Thus, 23 units is a lower bound based on this network's limitations. Similar situations arise when e_{19} and e_{25} are initiated first. If the initial transfer does not include a transfer connected to V_{20}, then V_{20} sits idle, so the overall time the schedule takes is greater than 23 time units. So the makespan for this network is 23 units, which agrees with our program's prediction.

Situation C

The network of 28 computers is expanded to include 42 transfers with varied transfer times and capacities. Computer V_{24}, which has five connecting transfers and the capacity to transfer only one file at a time, requires the maximum time to finish of all the computers; so a lower bound for the network to finish is the sum of V_{24}'s transfers, 29.6 units. The schedule produced by our program, shown in **Table 3**, takes 29.6 units, so this schedule runs at the makespan.

Table 3.

Schedule for situation C.

At time	Start transfers
0.0	$e_4, e_6, e_{10}, e_{11}, e_{12}, e_{13}, e_{15}, e_{17}, e_{18}, e_{20}, e_{24}, e_{26}, e_{35}, e_{41}$
2.1	e_{32}
3.0	e_9, e_{14}
4.4	e_{28}
5.0	e_{23}, e_{36}
6.1	e_{31}
6.2	e_{16}
7.0	$e_3, e_{19}, e_{33}, e_{38}$
8.0	e_1, e_{40}
9.0	e_8
10.2	e_{30}
11.0	e_{37}, e_{39}
14.0	e_{21}, e_{27}, e_{42}
14.2	e_{29}
15.1	e_5
17.3	e_2
20.1	e_7, e_{25}
23.0	e_{34}
25.4	e_{22}

Total time elapsed: 29.6 units

Strengths and Weaknesses

The greatest strength of our model is that the schedules produced run at the makespan for all three situations. This algorithm works well on all networks in which the computer that requires the maximum transfer time is capable of transferring only one file at a time. When this computer has the capacity of simultaneously transferring more than one file, the modified bin-packing algorithm used to find a lower bound starts to break down, as seen in the **Example** below. Even if we don't find the makespan, we get a schedule that runs in near-makespan time.

Our model is completely adaptable to other networks; and expansion or reduction of transfers, computers, or capacities is easily accommodated.

One weakness of our model is that all files are given equal priority (i.e., there is no capacity for a "rush transfer"). Each file transfer gets scheduled, but the transfers may not occur in a desired order.

This weakness goes hand in hand with the assumption that all computers contain the necessary information to complete their transfers. This may not be the case in all companies; e.g., computer B cannot transfer information to computer C until it receives its files from computer A. Our model does not take this type of network into account.

Another weakness of our model is the $O(m^3)$ running time.

Example. Say that the computer requiring the maximum transfer time also can simultaneously transfer three files. The eight connecting transfers and their times are shown in **Table 4**.

Table 4.

The eight connecting transfers and their times.

Edge	Transfer time
e_1	9.0
e_2	8.6
e_3	7.2
e_4	4.7
e_5	4.3
e_6	4.1
e_7	3.7
e_8	2.0

Using our modified bin-packing algorithm, e_1, e_2, and e_3 would all be placed in separate bins. Then

- e_4 would be placed in the same bin as e_3, summing to 11.9;
- e_5 would go in the bin with e_2, summing to 12.9;
- e_6 would go in the same bin as e_1, summing to 13.1;
- e_7 would go in the bin with e_3 and e_4, summing to 15.6; and

- e_8 would go in the bin with e_2 and e_5, summing to 14.9.

The picture would look something like **Table 5**.

Table 5.

Result after the modified bin-packing algorithm.

Bin 1	Bin 2	Bin 3
9.0	8.6	7.2
4.1	4.3	4.7
	2.0	3.7
13.1	14.9	15.6

By exchanging the 4.7 in bin 3 with the 4.1 in bin 1, the maximum time required by this computer would be 15.0 instead of 15.6.

Table 6.

Result after exchange.

Bin 1	Bin 2	Bin 3
9.0	8.6	7.2
4.7	4.3	4.1
	2.0	3.7
13.7	14.9	15.0

Sensitivity Testing

To test the algorithm, we considered changes in the situations:

- If the number of processes per computer is changed, the makespan would change accordingly. In situation C, when we reduced the capacity of the all the computers to 1, the makespan stayed exactly the same.

- If only the number of computers is changed, the algorithm would produce a schedule that runs in makespan or near-makespan time.

- If the number of transfers is increased or decreased, the algorithm again would adjust accordingly. In one example, however, an increase in the number of transfers did not affect the makespan: In situation B, when e_{28}, e_{29}, e_{30}, e_{31}, e_{32}, e_{36}, e_{38}, e_{39}, and e_{41} (as defined in situation C) were added to the network, the makespan stayed the same.

- Even if all of the computers in the network transfer at the same time, the makespan may be reduced if hardware is used efficiently. For example, we were able to reduce the makespan in situation C (see **Table 3**) by making a few computer exchanges. By simply exchanging V_2 and V_{24}, V_9 and V_{20}, V_1 and V_3, and V_8 and V_{22}, the makespan can be reduced from 29.6 to 19.3 units. This improves the makespan by 35%.

Recommendations

We recommend that the company run the match program before any actual transfers take place. With the output of the program, they should set up the network so that the transfers can occur exactly as suggested. This eliminates the delay in time required by the implementation of the matching program.

We also recommend that the company make the most efficient use of its computers by putting the computers capable of the most transfers in the locations that require the most transfers. This recommendation is made only under the assumption that all these computers run at the same speed.

References

Foulds, L.R. 1992. *Graph Theory Applications.* New York: Springer-Verlag.

Helman, P., R. Veroff, and F. Carrano. 1991. *Intermediate Problem Solving and Data Structures: Walls and Mirrors.* New York: Benjamin/Cummings.

Sedgwick, R. 1988. *Algorithms.* 2nd ed. New York: Addison-Wesley.

Wilson, R.J. 1985. *Introduction to Graph Theory.* 3rd ed. Essex, England: Longman Group Ltd.

Judge's Commentary: The Outstanding Communications Network Papers

Patrick J. Driscoll
Dept. of Mathematical Sciences
U.S. Military Academy
West Point, NY 10996–1786
driscoll@euler.math.usma.edu

The Communications Network Problem provided contestants with a fascinating example of how a real-world problem can submit to a range of mathematical analyses, spanning the simple to the sophisticated. It provided sufficient detail to clearly define the problem in both a numerical sense and a graph-theoretical sense and to appeal to the practical intuition of contestants. Based on the number of entries for this problem, these qualities apparently were not lost on the audience.

As in past years, the diverse backgrounds of the undergraduate contestants motivated many interesting modeling approaches, making the judging of these papers a formidable task indeed. Rather than simply characterizing the qualities of the outstanding papers and adding to the plethora of modeling checklists available in the literature, it is instructive to also include some comments regarding common shortcomings observed among the entries.

As many teams discovered, the numerical aspects of the Communications Network Problem could be solved directly using hand calculations. However, although a direct approach allowed many teams to obtain appropriate numerical results for this specific problem instance, in several cases it severely limited a team's ability to generalize their modeling approach to a larger class of networks, or to suggest possible improvement reconfigurations for the network. Consequently, their results possessed limited applicability to other communications network problems whose parameters varied from those presented.

By and large, the exceptional papers provided conclusive evidence that their teams had dedicated a substantial amount of time discussing and conceptualizing the problem prior to deciding on a modeling methodology. Rather than leaping head-first into an exhaustive literature search hoping to find a problem exactly like the one presented, they tended to characterize the problem in a more general sense and then identify several different modeling approaches to the problem. Lest this point be misinterpreted, it is not to say that "more is better" with regards to the number of models applied to this problem. The best papers appropriately recognized the time available for analysis, and presented a sufficiently complete treatment to convince the reader of their depth of understanding independent of the number of models that they used.

In contrast, papers that attempted a "partial-credit" or "core-dump" approach to their modeling effort, by briefly addressing many different models, detracted from the overall quality of their submission. It was far better to have presented one or two well-developed approaches than to dedicate one paragraph each to several different techniques with the sole intention of demonstrating an awareness of these techniques.

Nearly all of the papers attempted computer implementations of their algorithms in order to examine the algorithm's performance on larger problem instances. Several of the outstanding papers used their computer implementation to try to expose underlying structural properties of the problem in a more general context. One paper was able to characterize the distribution of the file-transfer times as a uniform distribution by using statistical tests. Given the number of readily accessible off-the-shelf computer simulation programs, it was important to identify exactly why a computer was being used to support analysis. There is little value in having run 10,000 simulations of a random network when there is no rationale given for the underlying probability distribution, the design of and motivation for the simulation, or the general structure of the network being simulated. Repetition is not the mother of invention, nor the supreme arbiter of scientific correctness.

The vast majority of superior papers recognized the applicability of Vizing's theorem and were able to develop specific graph-theoretical solution algorithms to take advantage of this fact. They also identified critical processors in the network that were limiting the optimal performance of the network file-transfer process. One paper was able to quantify cleverly the critical characteristic of these processors by developing a ranking structure for each node in the network, based upon layered sums of linked-computer capacities.

Several inferior papers presented modeling approaches so specifically tailored to this instance of the problem that they rendered generalization next to impossible. The better papers maintained a perspective that this problem was one instance in a class of problems. These papers sought first to identify the general class, next develop an algorithm addressing this larger class, and then narrow the scope of the algorithm to the instance at hand. These papers explicitly recognized the limitations of an enumeration-based approach for larger problems and attempted to develop more efficient approaches, hoping to avoid the computational burden imposed by the combinatorial nature of the problem.

Although the exceptional papers all expressed a healthy respect for the difficulty in attempting to identify an algorithm that works for all networks, several other papers repeatedly overstated their results, perhaps in response to the excitement of finding a useful algorithmic approach that supported their intuition. Claims to have found "the optimal solution"—without verifying uniqueness—or to extending results to "any arbitrary network"—after demonstrating a very restrictive applicability—diluted the credibility of their analysis. In spite of achieving logical and seemingly correct results, one must also recognize that optimality and uniqueness are not synonymous.

Lastly, the finer papers were consistently characterized by clear, logical, well-supported presentations that illuminated the team's underlying analytical reasoning. Those papers that were well-written had very few grammatical errors, conveyed their results in well-designed tables and graphics, and presented a complete summary typically made their point more effectively.

The Communications Network Problem was challenging, and many excellent solutions were offered. Five papers stood out from all the others, and those teams should feel proud of their accomplishment.

About the Author

Patrick J. Driscoll completed his undergraduate studies at the U.S. Military Academy at West Point in 1979, where he majored in engineering with a minor in mathematics. After serving in the infantry for eight years, Pat went on to earn a master's degree in operations research and a master's degree in engineering economic systems at Stanford University in 1989 and a Ph.D. in industrial and systems engineering at Virginia Tech in 1995. He has been a member of the faculty in the Dept. of Mathematical Sciences at West Point since 1989. His current research interests focus on discrete optimization in mathematical programming. Pat was an associate judge for the Communications Network Problem.

Participant's Commentary: Finding Makespans Is NP-Complete

Clifford A. McCarthy
Dept. of Mathematics
University of Illinois
Urbana, IL 61801
mccarthy@math.uiuc.edu

We consider a special case of the general optimization problem of finding the makespan for an arbitrary network—the case in which the graph of the network is a tree. We show that this more limited problem is polynomial-time reducible to an NP-complete problem, the partition problem.

Let G be a tree representing a network, with weights $T_1, T_2, \ldots, T_k$ assigned to its k edges. Interpret each edge-weight as the time required for a file transfer between the two nodes. Require both that no node can be involved in more than one transfer at a time and that transfers be *atomic* (they cannot be interrupted and resumed later).

If we can solve the *decision problem*:

Can all the transfers be performed in time n or less?

then we can also solve the *optimization problem*:

What is the minimum time to perform all of the transfers?

Starting with $n = 1$, we simply solve the decision problem for each successive value of n until we get a "yes" answer. We are guaranteed to reach a "yes" answer, since any graph can complete its transfers in $\sum T_i$ units of time by doing them one after another.

Theorem: *The decision problem is NP-complete.*

Proof: The proof involves two parts [Manber 1989, 341–357]. We must show that

- the decision problem is in the class NP, meaning that we can check in polynomial time whether a proposed solution is in fact a valid solution; and

- some NP-complete problem is polynomial-time reducible to the decision problem, meaning that we can convert (in polynomial time) any instance of the known NP-complete problem to an instance of the decision problem, such that the answer for the decision problem is positive if and only if the answer for the NP-complete problem is positive.

The first part of the proof is easy. Given an integer n and a proposed schedule for any graph, we can test in polynomial time whether it is a valid schedule requiring no more than time n, by checking each node to see if it is ever involved in two transfers, and checking if any transfers happen after time n. Hence, the graph scheduling problem (the decision version) is in the class NP.

For the second part, we use as our known NP-complete problem the *partition problem* [Garey and Johnson 1979, 60–62]:

Given integers $a_1, a_2, \ldots, a_m$, is there a partition of these integers into sets A and B so that
$$\sum_{a_i \in A} a_i = \sum_{a_i \in B} a_i ?$$

We first exhibit how to convert an instance of the partition problem into an instance of the decision problem. Let an instance of the partition problem be given, with integers $a_1, a_2, \ldots, a_m$. If $\sum a_i$ is odd, then there cannot be a partition into two sets with equal sum. So suppose $\sum a_i$ is even, with $\sum a_i = 2n$. Construct a tree with the structure and edge weights shown in **Figure 1**. Given an instance of the partition problem, we can certainly produce a description of the corresponding tree in polynomial time.

Finally, we show that *a partition exists if and only if the transfers on the tree can be performed in $(2n+1)$ units of time*. In other words, the partition problem is polynomial-time reducible to the decision problem, so the decision problem is NP-complete.

($\Longleftarrow$) Suppose that the transfers indicated in **Figure 1** can be performed in $(2n+1)$ units of time. This implies that the two nodes with edge weights n and $(n+1)$ spend the entire time engaged in transfers. Each can either perform the n-unit transfer followed immediately by the $(n+1)$, or the $(n+1)$ followed by the n.

If they both perform $(n+1)$-unit transfers first, then the two n-unit transfers will not be possible at the same time, because they are incident upon a common node. Hence, at least $(n+1) + n + n = 3n + 1$ units of time are required, which contradicts our supposition that the transfers can be performed in $(2n+1)$ units of time. Similarly, if both n-unit transfers are performed first, the $(n+1)$-unit transfers will not be possible at the same time, and $n + (n+1) + (n+1) = 3n + 2$ units of time will be necessary.

Hence, one of the $(n+1)$'s must be first, and the other must be last. This means that the only time that the 1-unit transfer can be performed is in the unit of free time between the two n's. During that transfer, the node connecting the a_i's is occupied (from time n to time $(n+1)$, assuming that the clock starts at time 0). Since $\sum a_i = 2n$, this node must also be

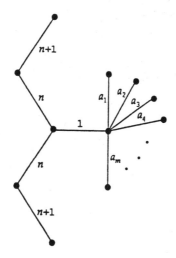

Figure 1. The tree representing the network of transfers.

continuously engaged in transfers. So the durations of the transfers that it performs before handling the 1-unit transfer must sum to n, and similarly for those performed after the 1-unit transfer.

Let A and B be the sets of the durations for the transfers handled in the two halves. Then A and B constitute a partition of $a_1, a_2, \ldots, a_m$, and

$$\sum_{a_i \in A} a_i = n = \sum_{a_i \in B} a_i.$$

If the transfers in the tree can be performed in $2n + 1$ units of time, there is a partition of $a_1, a_2, \ldots, a_m$ such that $\sum_{a_i \in A} a_i = \sum_{a_i \in B} b_i$.

($\Longrightarrow$) Now suppose that there is such a partition into sets A and B. As before, we can run the two $(n + 1)$-unit transfers, the two n-unit transfers, and the 1-unit transfer in a total of $(2n + 1)$ units of time, if the 1-unit transfer is run at the halfway point. The node joining the a_i's is then available for n units of time before this transfer and for n units of time after this transfer. We can handle the transfers corresponding to the elements of A in the first n-unit interval, and those corresponding to B in the last n-unit interval. So, if there is a partition of $a_1, a_2, \ldots, a_m$ into A and B such that $\sum_{a_i \in A} a_i = \sum_{a_i \in B} b_i$, then the transfers in the corresponding tree can be performed in $2n + 1$ units of time.

Hence, the desired partition exists if and only if the corresponding tree can be executed in $(2n + 1)$ units of time. $\qquad\square$

References

Garey, M.R. and D.S. Johnson. 1979. *Computers and Intractability: A Guide to the Theory of NP-Completeness*. San Francisco, CA: W.H. Freeman.

Manber, Udi. 1989. *Introduction to Algorithms: A Creative Approach*. Reading, MA: Addison-Wesley.

About the Author

Clifford McCarthy completed a mathematics B.S. in 1994 at Harvey Mudd College and is continuing in mathematics as a graduate student at the University of Illinois. His team's entry in the MCM, with fellow students Brian Diggs and Andrew M. Ross (advisor: David Bosley), was judged Meritorious. The proof in this commentary is his own; it constituted an appendix to their entry.

Other Problems of the First Ten Years

1985: The Animal Population Problem

Choose a fish or mammal for which appropriate data are available to model it accurately. Model the animal's natural interactions with its environment by expressing population levels of different groups in terms of the significant parameters of the environment. Then adjust the model to account for harvesting in a form consistent with the actual method by which the animal is harvested. Include any outside constraints imposed by food or space limitations that are supported by the data. Consider the value of the various quantities involved, the number harvested, and the population size itself, in order to devise a numerical quantity that represents the overall value of the harvest. Find a harvesting policy in terms of population size and time that optimizes the value of the harvest over a long period of time. Check that the policy optimizes this value over a realistic range of environmental conditions.

Comments by the Contest Director

The problem was contributed by Ervin Y. Rodin (Dept. of Systems Science and Mathematics, Washington University, St. Louis, MO). The animals chosen by the Outstanding teams were North American white-tailed deer (Harvey Mudd College), grizzly bears in Yellowstone Park (Mt. St. Mary's College), salmon (Southern Methodist University), and the Peruvian anchovy (Washington University). The Outstanding papers were published as follows:

Caulfield, Michael, John Kent, and Daniel McCaffrey. 1986. Harvesting a grizzly bear population. *College Mathematics Journal* 17 (1) (January 1987): 34–46.

Caulkins, Jonathan, Rob Marrett, and Andrew Yates. 1985. *The UMAP Journal* 6 (3) (1985): 27–49.

Special Issue: Mathematical Competition in Modeling. 1985. *Mathematical Modeling: An International Journal* 6 (6): 487–548. This special issue contains all four Outstanding papers.

1986: The Emergency-Facilities Location Problem

The township of Rio Rancho has hitherto not had its own emergency facilities. It has secured funds to erect two emergency facilities in 1986, each of which will combine ambulance, fire, and police services. **Figure 1** indicates the demand, or number of emergencies per square block, for 1985. The "L" region in the north is an obstacle, while the rectangle in the south is a part with a shallow pond. It takes an emergency vehicle an average of 15 seconds to go one block in the N–S direction and 20 seconds in the E–W direction. Your task it to locate the two facilities so as to minimize the total response time.

- Assume that the demand is concentrated at the center of the block and that the facilities will be located on corners.

- Assume that the demand is uniformly distributed on the streets bordering each block and that the facilities may be located anywhere on the streets.

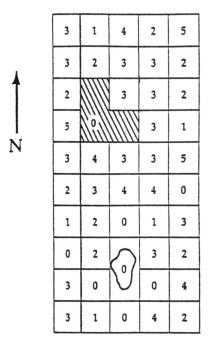

Figure 1. A map of Rio Rancho, with number of emergencies in 1985 indicated for each block.

Comments by the Contest Director

The problem was contributed by J.C. McGrew (Dept. of Geography and Regional Planning, Salisbury State College, Maryland). Originally, it had a stochastic aspect, but the Chief Judge (D.R. Morrison, Dept. of Computer Science, University of New Mexico) and I (Ben Fusaro) thought that it might be too much for three undergraduates on a weekend. How wrong we were! The approaches mainly took the form of an exhaustive search.

The Outstanding papers, by teams from Georgetown University, Harvey Mudd College, Grinnell College, and Washington University, together with commentaries, were published as follows:

Hardy, Maureen, Michael Irizarry, and Stephen Penrice. 1986. Help! To the rescue in Rio Rancho. *The UMAP Journal* 7 (4) (1986): 299–307.

Special Issue: Mathematical Competition in Modeling. 1986. *Mathematical Modeling: An International Journal* 7 (4): 595–652. This special issue contains all four Outstanding papers.

1987: The Parking Lot Problem

The owner of a paved, 100-ft-by-200-ft, corner parking lot in a New England town hires you to design the layout, that is, to design how the "lines are to be painted."

You realize that squeezing as many cars into the lot as possible leads to right-angle parking with the cars aligned side by side. However, inexperienced drivers have difficulty parking their cars this way, which can give rise to expensive insurance claims. To reduce the likelihood of damage to parked vehicles, the owner might then have to hire expert drivers for "valet parking." On the other hand, most drivers seem to have little difficulty in parking in one attempt if there is a large enough "turning radius" from the access lane. Of course, the wider the access lane, the fewer cars that can be accommodated in the lot, leading to less revenue for the parking lot owner.

Comments by the Contest Director

The problem was contributed by Maurice D. Weir (Mathematics Dept., Naval Postgraduate School, Monterey, CA).

Of the 156 papers entered in the Contest, 125 dealt with this problem. The Outstanding papers were by teams from Calvin College and Rensselaer Polytechnic Institute, both of which concluded that the lot could be designed to hold more than 75 average-sized cars. Their papers, together with commentaries, were published as follows:

Bingle, Richard, Dan Meindertsma, and William Oostendorp. 1988. Designing the optimal placement of spaces in a parking lot. *The UMAP Journal* 9 (1) (1988): 13–35.

Special Issue: Mathematical Competition in Modeling. 1987. *Mathematical Modeling: An International Journal* 9 (10): 765–784. This special issue contains all four Outstanding papers.

1988: The Railroad Flatcar Problem

Two railroad flatcars are to be loaded with seven types of packing crates. The crates have the same width and height but varying thickness (t, in cm) and weight (w, in kg). **Table 1** gives, for each crate, the thickness, weight, and number available. Each car has 10.2 meters of length available for packing the crates (like slices of toast) and can carry up to 40 metric tons. There is a special constraint on the total number of C_5, C_6, and C_7 crates because of a subsequent local trucking restriction: The total space (thickness) occupied by these crates must not exceed 302.7 cm. Load the two flatcars (see **Figure 1** so as to minimize the wasted floor space.

Table 1.

The thickness, weight, and number of each kind of crate.

	C_1	C_2	C_3	C_4	C_5	C_6	C_7	
t	48.7	52.0	61.3	72.0	48.7	52.0	64.0	cm
w	2,000	3,000	1,000	500	4,000	2,000	1,000	kg
	8	7	9	6	6	4	8	

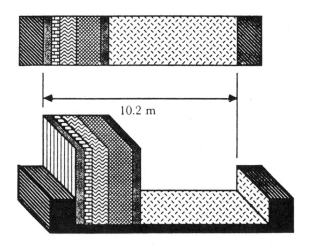

Figure 1. Diagram of loading of a flatcar.

Comments by the Contest Director

The problem was suggested by John J. Bartholdi III (School of Industrial and Systems Engineering, Georgia Institute of Technology). It is a modification of an unsolved problem that surfaced at the Ford Motor Company.

The Outstanding papers were by teams from Harvard University, University of California–Berkeley, University of Toronto, and the U.S. Military Academy. Their papers, together with a commentary by problem author Bartholdi, were published in *The UMAP Journal* 9 (4) (1988): 343–403.

1989: The Aircraft Queueing Problem

A common procedure at airports is to assign aircraft (A/C) to runways on a first-come-first-served basis. That is, as soon as an A/C is ready to leave the gate ("push back"), the pilot calls ground control and is added to the queue. Suppose that a control tower has access to a fast online database with the following information for each A/C:

- the time it is scheduled for pushback;

- the time it actually pushes back;

- the number of passengers on board;

- the number of passengers who are scheduled to make a connection at the next stop, as well as the time to make that connection; and

- the schedule time of arrival at its next stop.

Assume that there are seven types of A/C with passenger capacities varying from 100 to 400 in steps of 50. Develop and analyze a mathematical model that takes into account both the travelers' and airlines' satisfaction.

Comments by the Contest Director

The problem was contributed by J. Malkevitch (York College (CUNY), New York, NY) and myself (Ben Fusaro).

The Outstanding papers were by teams from Drake University, Harvey Mudd College, North Carolina School of Science and Mathematics, Ohio State University, and University of Dayton. Their papers, together with commentaries, were published in *The UMAP Journal* 10 (4) (1989): 343–415.

1990: The Snowplow Problem

The solid lines of the map (see **Figure 1**) represent paved two-lane county roads in a snow-removal district in Wicomico County, Maryland. The broken lines are state highways. After a snowfall, two plow-trucks are dispatched from a garage that is about 4 miles west of each of the two points (*) marked on the map. Find an efficient way to use two trucks to sweep snow from the county roads. The trucks may use the state highways to access the county roads.

Assume that the trucks neither break down nor get stuck and that the road intersections require no special plowing techniques.

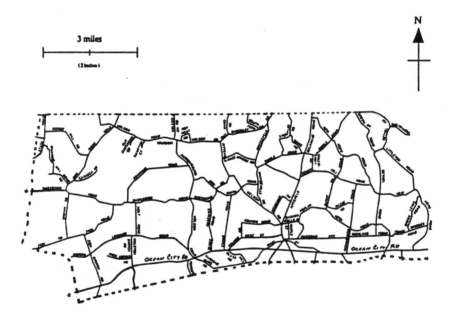

Figure 1. Roads in Wicomico County, MD.

Comments by the Contest Director

Most of the Outstanding teams characterized the problem as the construction of two Euler circuits with lengths as equal as possible.

The source of the problem was Kirk Banks, Roads Engineer and Head of Wicomico County Roads Division, Salisbury, MD. He had noticed that his crews took different times and wondered if there was a method that was superior to their "turn right" method. The county actually uses three snowplows, but the MCM problem statement modified this number to two because of the greater difficulty in solving and judging the problem with three plows. The MCM results show, however, that the students would have been able to handle three plows.

The Outstanding papers were by teams from Rose-Hulman Institute of Technology, Southern Oregon State University, U.S. Air Force Academy, and University of Alaska Fairbanks. Their papers, together with a commentary, were published in *The UMAP Journal* 11 (3) (1990): 231–274.

1991: The Water Tank Problem

Some state water-right agencies require from communities data on the rate of water use, in gallons per hour, and the total amount of water used each day. Many communities do not have equipment to measure the flow of water in or out of the municipal tank. Instead, they can measure only the *level* of water in the tank, within 0.5% accuracy, every hour. More importantly, whenever the level in the tank drops below some minimum level L, a pump fills the tank up to the maximum level, H; however, there is no measurement of the pump flow, either. Thus, one cannot readily relate the level in the tank to the amount of water used while the pump is working, which occurs once or twice per day, for a couple of hours each time.

Estimate the flow out of the tank $f(t)$ at all times, even when the pump is working, and estimate the total amount of water used during the day. **Table 1** gives real data, from an actual small town, for one day.

The table gives the time, in seconds, since the first measurement, and the level of water in the tank, in hundredths of a foot. For example, after 3316 seconds, the depth of water in the tank reached 31.10 feet. The tank is a vertical circular cylinder, with a height of 40 feet and a diameter of 57 feet. Usually, the pump starts filling the tank when the level drops to about 27.00 feet, and the pump stops when the level rises back to about 35.50 feet.

Table 1.

Water-tank levels over a single day for a small town. Time is in seconds and level is in 0.01 ft.

Time	Level	Time	Level	Time	Level
0	3175	35932	pump on	68535	2842
3316	3110	39332	pump on	71854	2767
6635	3054	39435	3550	75021	2697
10619	2994	43318	3445	79254	pump on
13937	2947	46636	3350	82649	pump on
17921	2892	49953	3260	85968	3475
21240	2850	53936	3167	89953	3397
25223	2797	57254	3087	93270	3340
28543	2752	60574	3012		
32284	2697	64554	2927		

Comments by the Contest Director

The problem was contributed by Yves Nievergelt (Mathematics Dept., Eastern Washington University, Cheney, WA). It is based on data from a consulting problem for Union, a town of 11,500 in northeastern Oregon. The Outstanding papers inspired immediate applications at the consulting firm, Equipment Technology and Design.

The Outstanding papers were by teams from Hiram College, Ripon College, and University of Alaska Fairbanks. Their papers, together with commentaries, were published in *The UMAP Journal* 12 (3) (1991): 201–241.

1992: The Air-Traffic-Control Radar Problem

You are to determine the power to be radiated by an air-traffic-control radar at a major metropolitan airport. The airport authority wants to minimize the power of the radar consistent with safety and cost.

The authority is constrained to operate with its existing antennae and receiver circuitry. The only option that they are considering is upgrading the transmitter circuits to make the radar more powerful.

The question that you are to answer is what power (in watts) must be released by the radar to ensure detection of standard passenger aircraft at a distance of 100 kilometers.

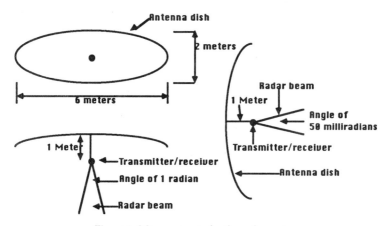

Figure 1. Measurements for the radar system.

Technical specifications (see also **Figure 1**):

- The radar antenna is a section of a paraboloid of revolution with focal length of 1 meter. Its projection onto a plane tangent to its vertex is an ellipse with a major axis of 6 meters and a minor axis of 2 meters. The main lobe energy beam pattern, located at the focus, is an elliptical cone that has a major axis of one radian and a minor axis of 50 milliradians. The antenna and beam are sketched in the figures provided below.

- The nominal class of aircraft is one that has an effective radar reflection cross-section of 75 square meters. For the purposes of this problem, this means that in your initial model, the aircraft is equivalent to a 100% reflective circular disc of 75 square meters, which is centered on the axis of the antennae and is perpendicular to it. You may want to consider alternatives or refinements to this initial model.

- The receiver circuits are sufficiently sensitive to process a return signal of 10 microwatts at the feed horn of the radar (which is located at the focus of the radar antenna).

Comments by the Contest Director

The problem was contributed by John Edwards (Dept. of Mathematical Sciences, U.S. Military Academy, West Point, NY). The problem statement contains several ambiguities and some unrealistic information, and a major factor in judging the papers was how the teams attempted to resolve these issues.

The Outstanding papers were by teams from Pomona College and University of Colorado–Boulder. Their papers, together with commentaries, were published in *The UMAP Journal* 13 (3) (1992): 205–225.

1993: The Optimal Composting Problem

An environmentally conscious institutional cafeteria is recycling customers' uneaten food into compost by means of microorganisms. Each day, the cafeteria blends the leftover food into a slurry, mixes the slurry with crisp salad wastes from the kitchen and a small amount of shredded newspaper, and feeds the resulting mixture to a culture of fungi and soil bacteria, which digest slurry, greens, and paper into usable compost. The crisp greens provide pockets of oxygen for the fungi culture, and the paper absorbs excess humidity. At times, however, the fungi culture appears unable or unwilling to digest as much of the leftovers as customers leave; the cafeteria does not blame the chef for the fungi culture's lack of appetite. Also, the cafeteria has received offers for the purchase of large quantities of its compost. Therefore, the cafeteria is investigating ways to increase its production of compost. Since it cannot yet afford to build a new composting facility, the cafeteria seeks methods to accelerate the fungi culture's activity, for instance, by optimizing the fungi culture's environment (currently held at about 120° F and 100% humidity), or by optimizing the composition of the mixture fed to the fungi culture, or both.

Determine whether any relation exists between the proportions of slurry, greens, and paper in the mixture fed to the fungi culture, and the rate at which the fungi culture composts the mixture. If no relation exists, state so. Otherwise, determine what proportions would accelerate the fungi culture's activity.

In addition to the technical report following the format prescribed in the contest instructions, provide a one-page nontechnical recommendation for implementation for the cafeteria manager.

Table 1 shows the composition of various mixtures in pounds of each ingredient kept in separate bins, and the time that it took the fungi culture to compost the mixtures, from the date fed to the date completely composted.